GRASS
NUTRITION

GRASS
NUTRITION

ROQUE RAMIREZ LOZANO

Library of Congress Control Number:		2015916209
ISBN:	Hardcover	978-1-5065-0808-5
	Softcover	978-1-5065-0809-2
	eBook	978-1-5065-0898-6

Print information available on the last page.

Rev. date: 29/09/2015

To order additional copies of this book, contact:
Palibrio
1663 Liberty Drive
Suite 200
Bloomington, IN 47403
Toll Free from the U.S.A 877.407.5847
Toll Free from Mexico 01.800.288.2243
Toll Free from Spain 900.866.949
From other International locations +1.812.671.9757
Fax: 01.812.355.1576
orders@palibrio.com
723532

CONTENTS

PRESENTATION

Grasses is the foremost plant type used for forage. For domesticated animals or wildlife, grass are the support of many individuals. This is due to the great number of grass types, their adaptability to wide habitats, and their persistence. Grass may be used to improve soil, diminish erosion, feed animals, absorb dung, create boundaries, clean air, disinfect water, offer habitat for wildlife including insects, defend waterways, and offer grain for humans. Recognizing what animals will require to be fed, tips to learning which grass will provide the best nutrition for better performance. Different animals have different nutritional requirements and diverse grasses effect animal performance in a different way. For example, lactating animals have high nutritional requirements and need high-quality forages; meanwhile, dry cows and recreational cattle may have dissimilar performance capacities and may have different rations.

This book examines in thirteen chapters, the nutritional characteristics of several cultivated and native grasses produced in northeastern Mexico and southern Texas, USA. It provides coverage of basic ruminant nutrition concepts. The author discusses the importance of grasses as food resource. He argues the nutrition of grass carbohydrates. This book covers research on silica and lignin content of grasses. The nutrition of grass proteins and grass digestibility is also emphasized. Details are given on intake of grasses. Importance is given to the fundamentals of grazing by ruminants. Wide coverage is presented on the nutritional role of trees and shrubs mixed with grasses. Contributions of the botanical and agricultural description of grasses grown in northeastern Mexico and southern Texas USA are discussed.

The Author

Prof. Roque Gonzalo Ramírez Lozano, Ph.D.
Universidad Autónoma de Nuevo León
Facultad de Ciencias Biológicas, Alimentos,
Ave. Pedro de Alba y Manuel Barragán S/N,
Ciudad Universitaria, San Nicolás de los Garza,
Nuevo León, 66455, México.
Mail: roque.ramirezlz@uanl.edu.mx

This book is dedicated to all my family for her kindness, and for her endless support to Emma my wife....

CHAPTER 1

Basic concepts of ruminant nutrition

Introduction

The digestive system of ruminants improves use of rumen fermentation products. This adaptation lets ruminants use resources (roughage) that may not be used by or are not available to other animals. Ruminants are in a unique position of being able to use such resources that are not in demand by humans but in turn provide man with a vital food source. Ruminants are also useful in converting vast renewable resources from pasture into other products for human use such as hides, fertilizer, and other inedible products (such as horns and bone). Ruminant livestock can use land for grazing that would otherwise not be suitable for crop production. Ruminant livestock production also complements crop production, because ruminants can use the byproducts of these crop systems that are not in demand for animals use or ingesting. Developing a good understanding of ruminant digestive anatomy and function can help livestock producers better plan appropriate nutritional programs and properly manage ruminant animals in various production systems.

Ruminant

Ruminants comprise are about 150 species, which include domestic (cattle, sheep and goats) and wild species (buffalo, deer, elk, giraffes, camels, etc.) that are found around the world. These animals all have a digestive system that is uniquely different from nonruminants (humans, pigs, poultry, dogs, etc.). They are able to obtain nutrients from edible plants-based feeds by fermenting them in a specialized stomach prior to digestion, principally through microbial activities. The process typically requires the fermented ingesta (identified as bolus) to be regurgitated and chewed again. The process of rechewing the bolus to further break down plant matter and stimulate digestion is named rumination.

Most ruminants belong to the suborder Ruminantia. Existing members of this suborder include the families Tragulidae (chevrotains), Moschidae (musk deer), Cervidae (deer), Giraffidae (giraffe and okapi), Antilocapridae (pronghorn), and Bovidae (cattle, goats, sheep, and antelope). Members of the Ruminantia suborder have a forestomach with four chambers. The nine existing species of chevrotain, also known as mouse deer and comprising the family Tragulidae, have four chambers, but the third is poorly developed. Chevrotains also have other features that are closer to nonruminants such as pigs. They do not have horns or antlers, and like the pigs, they have four toes on each foot.

Taxonomy of ruminants

Subclass: Ungulata
    Order: Artiodactyla
        Suborders
            -Ruminantia
                Families
                Tragulidae: Chevrotain, mouse deer
                Giraffidae: Giraffes
                Cervidae: Deer, moose
                Bovidae: Pronghorn, bison, buffalo, cattle, goats, sheep
            -Tylopoda
                Family
                Camelidae: Camels, Llamas

Although considered ruminants (any ungulate of the order Artiodactyla that chews its bolus) camelids differ from those members of Ruminantia in several ways. They have a three-chambered rather than a four-chambered digestive tract; an upper lip that is split in two with each part separately mobile; an isolated incisor in the upper jaw; and, uniquely among mammals, elliptical red blood cells and a special type of antibodies lacking the light chain, besides the normal antibodies found in other species.

Ruminant gastrointestinal track

Ruminants have one stomach with four compartments (Figure 1.1). The four parts of the stomach are the rumen, reticulum, omasum, and abomasum. In the first two chambers, the rumen and the reticulum, the feed ingesta is mixed with saliva and separates into layers of solid and liquid material. From 60 to 75% of ingesta fermented by microbes before exposed to gastric juices in the abomasum. Solids clump together to form the bolus. The rumen is the largest section of the four compartments and the foremost digestive center. The bolus is then regurgitated and chewed to completely mix it with saliva and to break down the particle size. Fiber, especially cellulose and hemi-cellulose, is primarily broken down in these chambers by microbes (mostly bacteria, as well as some protozoa, fungi and yeast) into the three main volatile fatty acids (VFA): acetic acid, propionic acid and butyric acid. Proteins and nonstructural carbohydrate are also fermented.

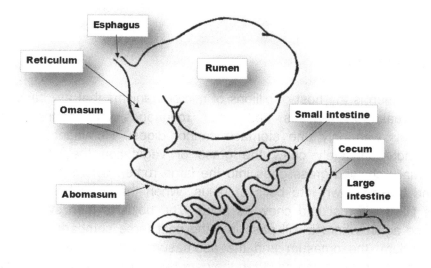

Figure 1.1. Schematic illustration of the gastrointestinal track of a cow

Classification of ruminants by feeding preferences

1. *Concentrate selectors*
   a. The properties are: evolved early, small rumens, poorly developed omasum, large livers and limited ability to digest fiber
      i. Classes are: 1) fruit and forage selectors; very selective feeders, examples: duikers and unis and 2) tree and shrub browsers; eat highly lignified plant tissues to extract cell solubles, examples: deer, giraffes, kudus
2. *Intermediate feeders*
   a. The properties are: seasonally adaptive,
   b. Feeding preference, prefer browsing, examples are: moose, goats, elands
   c. Prefer grazing, examples are: sheep, impalas
3. *Roughage grazers*
   a. The properties are: most recently evolved, larger rumens and longer retention times, less selective and digests fermentable cell wall carbohydrates
   b. The classes are: 1) fresh grass grazers, examples buffalo, cattle, gnus, 2) roughage grazers, examples hartebeests, topis and 3) dry region grazers, example camels, antelope, oryxes

Rumination

The rumen is the host of billions of microorganisms that are capable to breakdown grasses and other roughages that nonruminant animals with only one stomach cannot digest. Ruminant animals do not completely chew the grass or vegetation they eat. The partially chewed grass goes into the large rumen where it is stored and broken down into balls of bolus. When the animal has eaten its fill, it will rest and chew its bolus. The bolus is then swallowed once again where it will pass into the next three compartments the reticulum, the omasum and the abomasum (true stomach). Cattle have a four-part stomach when they are born. However, they function primarily as a nonruminant (simple-stomached) animal during the first part of their lives. At birth, the first three compartments of a stomach of a calf (rumen, reticulum, and omasum) are inactive and immature. As the calf grows and begins to eat a variety of feeds, its stomach compartments also begin to grow and modify. The abomasum constitutes nearly 60 percent of the young stomach of a calf, decreasing to about 8 percent in the mature cow. The rumen includes about 25 percent of the young stomach of a calf, growing to 80 percent in the mature cow. Ruminants have the ability to convert the plants and crop residues into high quality protein in the form of meat and milk. Moreover, they feed on the discards and cutting from fruit and vegetable farming and the byproducts from food processing industries.

Digestion

Digestive enzymes carry out process by which proteins, fats, and carbohydrates are broken down into absorbable molecules. To obtain forages, the ruminant animal utilize its mouth and tongue to cut plants when grazing or intake collected foods. Ruminants select forages through grazing when wrapping their tongues round the grass and at that time heaving to rip the plants for intake. It seems that cattle uses a range of 25,000 to 40,000 bites daily to cut forage during grazing. The ruminant employ of all day, a-30 percent period foraging, a-30 percent period ruminating, and little less than a-30 percent period wasting when animals are resting.

There are not incisors teeth in the top of the ruminant mouth that is a palate lax and firm. The bottom incisors teeth pressure contrary to the firm dental pad. The incisors teeth of ruminants that select roughage and grass are extensive with a tool-cut crest; meanwhile, the concentrate selectors are finer and shape. Molars and premolars teeth tie between superior and inferior jaws. They are teeth that press and grind selected plants when chewing and rumination is initiated.

There are different types of glands that secret saliva (parotid, molars, buccal, lingual, sublingual, submandibular, lip, and throat) but they can be classified according to the type of saliva secretion. The mucilaginous secretion aims to diminish the bolus and facilitate chewing and swallowing while alkaline saliva, especially formed by carbonates, bicarbonates and phosphate maintains the pH in the rumen, near neutral narrow range, and acts the same time as the bicarbonate is usually taken to avoid stomachache. Furthermore, saliva, which contains urea, keeps a level of more or less constant nitrogen in the rumen. Salivary secretion ruminants is very abundant and variable. It is estimated that in cattle between 90 and 190 liters per day, according to various authors and various diets.

Considerable amounts of animal food ingesta (less than 3.5 cm) quickly is consumed and swallowing lacking of mastication. In ruminants, esophagus roles in two directions, permitting the animals to regurgitate their bolus for further mastication, when is required. The rumination process or mastication the bolus is where ingesta and other forages are obligated back to the mouth for more mastication and mixed with the saliva. The bolus is at that time swallowed another time and delivered to the reticulum. Then, the hard ingesta is slowly moved to fermentation into the rumen; meanwhile, utmost of the liquid fraction quickly is moved from the reticulorumen to the omasum and eventually to the abomasum. In the rumen, the solid left ingesta usually remnants for maximum to 48 h and it forms a solid floorcovering in the rumen, where microbial organisms may utilize the fibrous parts to create energy precursors.

Because of the rumen and reticulum have comparable purposes, both are intentional named as rumenreticulum and are divided just by a slight muscular doubling of tissular material. The principal feature of the reticulum is to gather shorter digested materials and transported them to the omasum; meanwhile the greater materials are kept in the rumen for additional breakdown. The rumen accomplishes as a fermentation container by holding fermentation carried out by microorganisms. A range of 50 to 65 percent of soluble sugars and starch ingested are processed in the rumen. In the omasum the water absorption is occurred. The real stomach of the ruminant is considered the abomasum. In the abomasum. the HCl and digestive enzymes, such as pepsin, are produced. The enzymes used in digestive processes produced by the pancreas, are delivered to the abomasum. All secreted compounds assistance to fix proteins to be absorbed into intestines. The abomasum pH varies from 3.5 to 4.0.

Both intestines, small and large are the sites of nutrient absorption of ingesta digested in the abomasum. In the small intestine where digesta is entered mixed with the secretory substances from liver and pancreas. In this site, the pH is elevated from 2.5 to a range of 7 to 8. Elevated pH is required for the enzymes to perform accurately. Into the duodenum, bile from the gall bladder is secreted. The bile helps in digestion process. The nutrient absorption occurs throughout the small intestine as an active process, in which, rumen by-pass protein absorption is considered.

The large intestine absorbs the water, and the remaining material then is excreted as feces throughout the rectum. At the commencement of the large intestine, the cecum is found and is a big blind bag. The colon is a part of the large intestine and is the place of mostly of the water is absorbed.

Fermentation

The forestomach of the ruminant and large intestine of caudal fermenters are outstanding, constant movement fermentation arrangements comprising great number of microorganisms. The microbes that digest structural carbohydrates and other molecules also compromise at most three other principal features:

1. Production of great superiority protein in the form of microorganisms. However, caudal fermenters might have not use gain of this action; but, in ruminants, bacterial and protozoal microbes are continuously fluid to the abomasum and then to the lower track, where they are processed and assimilated. All ruminants need definite type amino acids, in which their tissues might have not manufactured (for example, indispensable amino acids). During fermentation, the microorganisms may create all the amino acids, and by this way, are delivered to the animal´s host.

2. Production of crude protein (CP) from NPN sources. Microorganisms might have use urea to produce CP. Certainly; ruminants usually are fed urea as a low-cost nutritional complement. In addition, ruminants, into saliva, secrete urea

performed during protein metabolism that moves to the rumen and aids as other nitrogen source for the microorganism.
3. Synthesis of B vitamins and vitamin K. Rumen microorganisms can produce all the B vitamins; thus, deficiency of one of them is difficult to find.

By rare exclusions, all soluble and structural carbohydrates and all kind of proteins can be used as substrates for rumen microbial fermentation. The cellulose account for 40 to 50% of most of stem, leaf and root of grasses. These fibers of cellulose are entrenched in a core of hemicelluloses and lignin compounds that are covalently linked. The bacteria and protozoa in the rumen or hindgut produce all the enzymes necessary to digest cellulose and hemicellulose. The free glucose from this procedure is thus occupied and break down by microbes, and the discarded products of bacterial breakdown are transported to the animal´s host. The starch is metabolized similarity.

Rumen pH characteristically varies from 6.5 to 6.8. The rumen environment is anaerobic (without oxygen). Gases produced in the rumen include carbon dioxide, methane, and hydrogen sulfide. The gas fraction rises to the top of the rumen above the liquid fraction. The main VFA are acetic acid, propionic acid and butyric acid, which together deliver for the majority of energy requirements for ruminants. The main VFA produced is always acetate. Animals feeding diets high in fiber, the molar proportion of acetic acid to propionic acid to butyric acid is about 70:20:10. The three-major VFA absorbed from the rumen have to some extent different metabolic destinies:

1. The acetic acid is used marginally in the liver, where is oxidized throughout most of the body tissues to produce energy in the form of ATP. Another important use is as the main source of acetyl CoA for lipogenesis.
2. Most the propionic acid is nearly completely removed from portal blood by the liver where propionate functions as a main substrate for gluconeogenesis, which is critical to the ruminant due to almost no glucose, enters the small intestine for absorption.

3. Most of the butyric acid that originates from the rumen as the ketone beta-hydroxybutyric acid is oxidized in body tissues for making energy sources.

Proteins play very important roles in almost all body processes that are related to 1) catalysis, 2) enzymatic, 3) control of metabolism, 4) immunology, 5) mechanical support 6) motion, 7) storage, and 8) transport. All proteins in the body tissues are in a state of continuous flux, and the size of the body protein pool is dependent on a balance between hydrolysis and synthesis.

In ruminants, all dietary proteins go into the rumen (Figure 1.2). Rumen microbial enzymes (proteases and peptidases) digest the majority of these proteins. Delivery peptides and amino acids are taken up by microorganisms, and utilized in several manners; as well as synthesis microbial protein. Nevertheless, a great amount of amino acids ingested by rumen microbes is deaminated and some follow to the same pathways utilized for the metabolism of carbohydrates. Thus, the result is that abundant of protein is metabolized and converted to VFA. The main products from the rumen fermentation are the VFA, which eventually are used in several ways, being the supreme importance of the VFA to ruminants is that they are absorbed and function as energy for the animal productivity.

The manner of how lipids are involved with ruminal fermentation is a multifaceted mechanism concerning to 1) partitioning of lipid into the membrane of the microbial cell, 2) strength of the lipid to interrupt membrane and function of the cell, 3) physical addition of microbial cells to plant surfaces, and 4) manifestation and action of hydrolytic enzymes of microorganisms. Two important microbial transformations of lipids in the rumen occur (Figure 1.3): lipolysis and hydrogenation. Lipolysis origins the relief of free fatty acids from esterified plant lipids followed by hydrogenation, which diminishes the amount of double bonds. However, the loss of fatty acids from the rumen both by absorption across the ruminal wall or by catabolism to VFA or CO_2 was minimal. Moreover, microbes are capable to new synthesize fatty acids from the precursors of carbohydrate. Hence, lipids reaching the duodenum are from fatty acids from either dietary or microbial origin. Rates of Lipolysis and hydrogenation differ depending of forage quality (e.g., maturity stage and N content), surface area of particles feed ingesta, and mechanical alterations of the lipid molecule that prevent attack by bacterial isomerases.

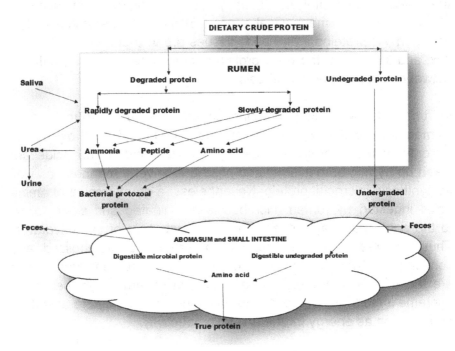

Figure 1.2. Schematic diagram of degradation of
dietary protein in the gastrointestinal track

It appears that lipids added to ruminant rations may importantly upset fermentation metabolism in the rumen, producing diminished digestion of nonlipid energy sources. It has been found that ruminal digestibility of structural carbohydrates can be reduced 50% or more by less than 10% of the dietary fat. In addition, the reduction in digestibility is complemented by diminished production of CH_4, H, and VFA, with a reduced acetate to propionate ratio. If fat supplements inhibit ruminal fermentation, will be limited hindgut fermentation may be less reduction of fiber digestibility in the whole gastro intestinal tract; however, excretion of fiber in feces often still occurs.

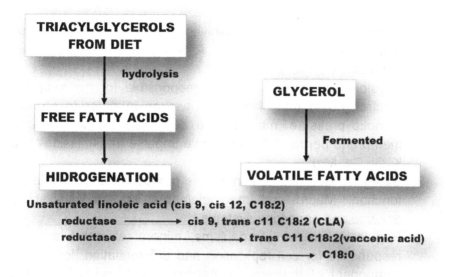

Figure 1.3. Diagram of ruminal lipid metabolism

Functions of minerals in the ruminant body

Minerals are inorganic elements found in small amounts in the ruminant body. Inorganic means that the substance does not contain carbon. The minerals found in body tissues and fluids of adult ruminants are originated mainly from exogenous sources and constitute approximately 4% of the body weight of the animal. Minerals are required for the normal functioning of all metabolic processes in ruminants. Dietary deficiencies or excesses of certain minerals may produce in economic losses in animal productivity. Minerals can be divided into macroelements (contents higher than 50 mg kg^{-1} of body weight) and trace elements or microelements (below 50 mg kg^{-1}). There is a list of 22 essential minerals for animal that comprise 7 macrominerals (calcium, chloride, magnesium phosphorus, potassium, sodium, and sulphur), and 15 trace elements (arsenic chromium, cobalt, copper, fluoride, iodine, iron, manganese, molybdenum, nickel, selenium, silicon, tin, vanadium and zinc).

Macro and microminerals play four key roles:

1. Structural. This function involves elements that build organ and tissue structures (Ca, Mg, P, Si in bones and teeth, P and S in muscle proteins).
2. Physiological. This function is responsible for the supply of electrolytes to body fluids and tissues in order to regulate osmotic pressure, maintain the acid-base balance, regulate membrane permeability and nerve impulse transmissions (Na, K, Cl, Ca, Mg).
3. Catalytic. This role of minerals is probably the most relevant function. Macro and macroelements act as catalysts in enzyme and endocrine systems.
4. Regulatory. In living organisms, mineral elements are also responsible for cell replication and differentiation. Iodine is a component of thyroxine, a hormone responsible for thyroid function and energy processes.

Mineral deficiency in ruminants may damage or even more inhibit metabolic paths necessary for normal function of the body, and may produce clinical symptoms. Severe macro or microelement deficiencies are showed by symptoms corresponding to the function of the deficient mineral in the body; therefore, contributing to an accurate diagnosis of the health problem. In a minor deficiency, the symptoms are nonspecific, often passing and difficult to diagnose because of the low intensity.

Functions of vitamins in the ruminant body

Vitamins are organic nutrients needed in small quantities to perform specific functions. They do not provide energy; however, are required in the use of energy. Vitamins aid the ruminant in regulating the body functions, protecting the body healthy, and helping resistance to illnesses. The deficiency of a vitamin can produce clinical symptoms of disease or death. Vitamins are classified as fat-soluble and water-soluble.

Fat-Soluble vitamins are those vitamins stored in the body fat and released, as the body tissues require them. These vitamins can be stored for prolonged periods. The vitamins are: A, D, E, and K. In general, the functions Vitamin A are related to maintaining internal and external coatings and is necessary in reproductive system of the ruminant. The plant pigment carotenes are the precursors of vitamin A; the body transforms carotenes into vitamin A. The deficiency symptoms of vitamin A comprise watery eyes, a rough hair coat, and low animal growth. Vitamin D regulates the absorption of Ca and P. Ruminants can synthetize their own vitamin D when are exposed to the sunlight. Vitamin E promotes good health protecting the ruminant body from lipid peroxidation. A lack of vitamin E causes failure in the reproductive system. Vitamin K is important in blood clotting. Vitamin K is not necessary in the diet, as microbes in the gastrointestinal track can synthetize it.

The water-soluble vitamins are synthetized by microorganisms in the rumen. Water-soluble vitamins include Vitamin C and the B vitamins. Vitamin C is synthesized in the body tissues. Therefore, it is not necessary to add it to feed rations. The B vitamins are categorized into two groups. In the group one are included: thiamin, riboflavin, niacin, and pantothenic acid. In general, these vitamins are involved in the release of energy from the nutrients of foodstuffs. In this group, two are included: folic acid and vitamin B12. In general, these vitamins regulate the synthesis of leucocytes.

Functions of water in the ruminant body

Water is the main constituent of the ruminant animal, the amount of water in the body varies from 50 to 80% of the live weight, depending on age and amount of body fat. A ruminant can lose nearly all of its fat and about 50% of its tissue protein may live. However, the loss of 10% of its body water might be lethal. A worthy water source of water may be defined in terms of both, quantity and quality of the water. It has been established that a worthy water source is relevant to the livestock management due to the total water intake is positively related to feed dry matter intake.

The four main functions of water in the ruminant body are: 1) helping in eliminate waste products of digestion and metabolism (feces of healthy cattle often contain 75 to 85% water), 2) regulating blood osmotic pressure, 3) a major component of secretions (milk and saliva) and 4) in the products of conception and growth in thermoregulation of the body as affected by evaporation of water from the respiratory tract and from the surface of skin. Cattle accomplish their requirements for water from three main sources:

1. Free drinking water or from snow
2. Water contained in feed
3. Metabolic water produced by metabolic reactions of nutrients

The first two sources are very important in livestock management, due to the high variations in water consumptions, an estimate of water consumptions of cattle should be made based on production features, which may affect water intake. Water intake requirements depend on the following features:

1. Air temperature surrounding to the ruminant
2. Amount of milk produced
3. Amount of weight gain
4. Bred and dimension of animal
5. Kind of diet
6. Level of dry matter intake
7. Pregnancy
8. Class and quality of water
9. Type and level of animal activity
10. Temperature of the water accessible.

CHAPTER 2

Features of grasses

Introduction

The Gramineae (Poaceae), is the family of grasses, cereals, bamboo and sugar cane, consist of about 650-700 genera and about 12,000 species spread throughout the world, even in the coldest or torrid regions, and are often dominant in major vegetation such as savannas, steppes and aquatic vegetation. The grasses are flowering plants belonging to the monocotyledon group (subclass Liliidae) with the embryo developed from a single sheet that is typically pollinated by the wind. Many of grasses are manly used as food for animals and humans, since this family belong the wheat (*Triticum* sp), barley (*Hordeum vulgare* L.), rice (*Oryza sativa* L.), corn (*Zea mays* L.), rye (*Secale cereale* L.), oats (*Avena* sp.) and other cereals, that stored in their fruits large amounts of carbohydrates (starch) and in lesser proportion fats and proteins. From *Saccharum officinarum* L. that is rich in sugars, it is obtained the sugar cane.

Phenology and forage source of grasses

The flowering plants are divided into two groups, the Monocotyledons and the Dicotyledons. Grasses, sedges and rushes all belong to the Subclass Monocotyledons. They can easily be distinguished from the Dicotyledons in that they have:

- One seed leaf.
- Leaves with parallel veins.

- Fibrous roots.
- Scattered vascular bundles (conducting tissue within the stems).

The grasses and the sedges belong to the Order Graminales (a further subdivision of the Monocotyledons). The flowers are inconspicuous, are arranged in spikelets and enclosed in chaffey (papery) scales. They do not have petals and sepals (image of a grass flowering head here). The fruit is one seeded. Its seed coat is united with the ovary wall and is called a Caryopsis. Grasses are in the Family Poaceae, (also known as Gramineae), and Sedges are in the Family Cyperaceae.

Grass stems are mainly cylindrical, elliptical in cross-section, articulated, called ordinarily culms, usually with solid nodes and hollow (but can be completely solid as in the case of maize and some bamboos). The nods are somewhat thicker than they are born inter nods, leaves, and sprouts. Internodes are sometimes somewhat flattened in the area where the branches are developed. A little further up there is a meristem knot ring-shaped interlayer determining stem elongation.

Leaves have alternate arrangement, typically composed of sheath, ligule and auricle. The sheath surrounds tightly to stem its margins overlap but do not fuse together (only occasionally can be found forming a tube). The ligule is a small membranous appendage, or rarely a group of hairs, located at the junction of ligule with the sheath, in the axial part. Leaf sheath is simple, usually linear, with parallel venation. Can be flattened or sometimes rolled into a tube, it may be continuous with the sheath or possess petiole.

The elemental flower of grass is a small pin formed by one or more flowers or sessile seated on an articulated spine, often very short, called rachilla protected by sterile bracts called glumes. The flowers can be hermaphrodite or unisexual and have a rudimentary spikelet two or three pieces. These are the organs that, to be turgid, anthecium determine the opening or during bloom floral box, allowing feathery stigmas and stamens are exposed. The florets are formed by the lemma, adhered to rachilla pálea and inserted on the floral axis on raquilla born in the armpit of the lemma and supports to the floral organs themselves. The lemma (or lower glume) is shaped keel, it can be mutica or awned and embraces palea their edges. The edge is born at the end of the lemma or on its back. The palea is lanceolate or higher glume cover enclosing the flower. Glumes are inserted on rachilla one lower than the other. All these elements are very variable so it is convenient to analyze them separately.

Growth and development of grasses

Because of grass species (Figure 2.1) have dissimilar growing periods, are recognized as annuals, biennials and perennials. The annual type, perish each year after they have created the seeds; meanwhile, the biennial type, its period is prolonged by two years, the perennial type yield both vegetative and flowers for several years.

Grasses are growing in dissimilar manners. There are three developing ways:

- Bunch Type (Caespitose), it crops bunches of grass.
- Stoloniferous, this produces stolons that path on surface of the soil. this has regular green leaves and root in the nodes of stems.
- Rhizomatous, rhizomes grow below the surface of grown and the plant is small white leaves.

A rhizome is considered as a stem of a plant that typically is encountered under the soil surface. The majority of grass species have two groups of chromosomes per each cell, which means that are diploid. However, the grass *Lolium perenne* is considered tetraploid because it has four groups of chromosomes in each cell. Tetraploid grasses have a larger cell size, wider leaves and scarcer tillers. Ruminants grazing this type of grasses experience larger dry matter intakes because these grasses have lesser structural carbohydrates.

The seed of the grass is named a caryopsis, which comprises an embryo and the endosperm considered a stock of starch. From the embryo is composed the first shoot (or plumule), the root (or radicle) and a scutellum (the first leaves). The germination comprises the endorsement of the water by the seed and this motivates breathing, division of cells and enzyme excretion. The enzymes act for the hydrolysis the starch to soluble sugars in plant endosperm. The sugars travel to the embryo to sustain growth of plumule and radicle.

The source of the above soil parts of the grass are found in the tip of the apex, which is a region named the apical meristem. In the apex also are found nodes that are the sockets of addition of leaves, which are divided by internodes. The internodes are considered part of the stem tissues that dispersed among nodes.

Once the apical meristem yields the leaves, it too yields a bud, which may grow into another tiller. These buds might have also develop to rhizomes or stolons that are significant stock tissues beneficial for development of the grass. In each grass leaf is developed a blade connected to a sheath that ambiances the stem over the node. At the time when the plant grows starting from the bottom, old leaves are pushed up. A continuous turnover of leaves per tiller are produced during the life of the grass due to the amount of leaves keeps continuous throughout the life of the grass.

The length of each day as well as the temperature stimulates the reproductive growth of grass plants. In the propagative period, the internodes of stems are extended. This may provokes a fast elongation of stems that lifts the tip of the shoot over ground. The flowers (that are the multiplicative arrangements) grows after the tip of the shoot apex.

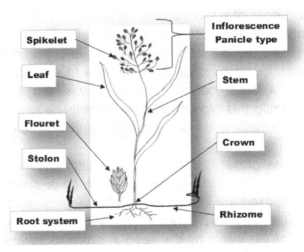

Figure 2.1. Basic Structure of a grass plant

Native grasses

Native grasses are the dominant plants in the vast range-land plant communities and are important in the understory of many forested areas. Fibrous roots of grasses hold the soil in place and build up soil fertility. Grasses, like all green plants, convert the sun's energy into carbohydrates for their own food and for use by animals and man. Native grasses are by definition those grasses that are indigenous (native) to a local region. This is in contrast to those grass species that have their origins in Europe, Asia, Africa or other parts of the world.

Over the time of white settlement in America and the pastoral development of the region, native grasses have largely been denigrated and replaced by other species. Many of the introduced species are considered to be more nutritious and hence have been accepted and used extensively. Unfortunately, those native grasses that have survived have often been the non-productive and less palatable species, giving them a negative reputation.

However, many of the native grass species are drought-resistant and require low input costs. These characteristics make them very suitable for inclusion in a balanced and sustainable grazing system, particularly in low rainfall areas.

Many native grasses are well adapted to surviving the heat and lack of moisture typical of many areas of the world (Table 1). These characteristics include special growth characteristics, such as thickened stems at the base for food storage and a corky integument over the roots to protect them against excessive heat. Some species have a different leaf structure from those species adapted to the higher rainfall areas. This leaf structure gives the plant small, thickened leaves or bristle-like leaves that help to reduce the amount of moisture that is transpired from the leaf surfaces.

Photosynthesis

The basic reaction in green plants that converts solar energy to chemical energy is called photosynthesis. This reaction is directly or indirectly responsible for all life on earth. It provides the energy (carbohydrate) for plant growth and maintenance as well as animal growth and maintenance.

Sunlight

$CO_2 + H_2O$ -------------> Carbohydrates (Food) + O_2 ---> Green plant material

The carbon of CO_2 (a gas) is converted to the carbon of the carbohydrate (a solid). The carbohydrate is a chemical way to store the sun's energy as food. Carbon dioxide in the air, a raw material for the photosynthetic process, is not very abundant. The atmosphere has approximately 0.03%. The success of a plant will depend on its ability to collect and use CO_2 in the photosynthetic process. For perennial pasture grasses to remain productive, the photosynthetic process must first feed the plant before it can provide feed for livestock.

Table 1. Genus and agronomic characteristics of native grasses grown in America.

Genus	Agronomic characteristics
Aristida	Hard, wiry and generally unpalatable, especially when in flower. Seed causes problems in wool. Crude protein: 2.0 - 13.8%
Bothriochloa	Exists on a wide variety of soil types with a preference for wetter areas. Forage value declines after flowering. Crude protein: 4.1 - 15%
Bouteloua	Produces high-quality nutritious forage that is readily eaten by livestock. Wild turkeys eat the seed.
Chloris	Generally of moderate value. Grazed readily prior to flowering. Crude protein: 4.0 - 10.9%
Dichanthium	Highly palatable and well regarded when green. Crude protein: 2.3 - 10.2%
Digitaria	Generally palatable even when dry. Crude protein: 3.7 - 16.8%
Eragrostis	Generally, poor grazing value; however will provide some dry feed and ground cover during drought.
Hilaria	It is grazed all year by horses, cattle, sheep, goats, antelope, and deer.
Leptochloa	All livestock graze it readily, especially when green and succulent. During dormant season, it furnishes good quality forage but should be supplemented with a protein concentrate.
Panicum	Grazing value moderate, new green growth particularly palatable especially prior to seeding.
Paspalum	Most of paspalums are fair forage for cattle and deer during spring and summer. Their value as winter roughage is poor. Quail and other birds eat the seed.
Setaria	Livestock graze it moderately, usually during spring and summer. It becomes unpalatable in fall and provides poor forage after maturity. When grazed for roughage, it should be supplemented with a mineral and protein concentrate.
Sporobolus	Produces abundant forage, which is grazed by cattle and horses.
Tridens	Cattle and horses graze it. Rodents and birds eat the seed.

Cool- (C3) and warm-season (C4) grasses

The C4 photosynthetic system found in warm season grasses is more efficient in gathering CO_2 than the C3 system. Consequently, warm-season plants have the potential to be more efficient than cool-season plants when both are at optimum conditions. Optimum temperature for the growth of C3 plants is around 18-24°C while it is 32-35°C for C4 plants. C4 plants may use up to 50% less water to produce a unit of dry matter. Water use efficiency and temperature optimums explain why warm-season pastures are more productive in hot, dry summer months and cool-season pastures are more productive in the cool, moist spring and fall months. Since cool-season plants start growth early in the spring, soil moisture is often depleted by early summer. In contrast, warm-season grasses that start their growth in late spring generally have a favorable soil moisture profile in early summer.

There is no all-season plant available. Producers must recognize the limitations of plant seasonality and take advantage of the variation in grasses. Incorporate both cool and warm-season pastures to provide a longer grazing in climates with a hot period during the summer.

C4 plants are also more efficient in nitrogen utilization. Warm-season plants recover more N from a given amount of available soil nitrogen than the cool-season plants due to increased soil microbial activity in the summer. In contrast, cool-season grasses have a high demand for nitrogen in the spring when there is little soil microbial activity. Consequently, nitrogen fertilization is essential to achieve satisfactory levels of production for cool-season grasses. Warm-season grasses generally respond better to fertilizer under humid climates.

These physiological factors help explain why cool-season grasses grow in the spring, mature by late spring or early summer, and become dormant during the hot summer months before resuming growth in the fall. Warm-season grasses mature during late summer and may become dormant early in the fall.

These physiological factors also have an effect on animal production. For example, the poor performance of livestock grazing tall fescue during the summer has been attributed to an endophyte fungus. Normal reduction in cool-season growth and quality during the summer is an important factor in grass utilization. The use of an endophyte-free, cool-season grass will only partially solve the problem of "summer slump." Matching the seasonality of grasses with the season of livestock use is a good strategy.

Examples of C3 plants:

Most small seeded cereal crops such as rice (Oryza sativa), wheat (Triticum spp.), barley (Hordeum vulgare), rye (Secale cereale), and oat (Avena sativa); soybean (Gycine max), peanut (Arachis hypogaea), cotton (Gossypium spp.), sugar beets (Beta vulgaris), tobacco (Nicotiana tabacum), spinach (Spinacea oleracea), potato (Solanum tuberosum); most trees and lawn grasses such as rye, fescue, and kentucky bluegrass. Also are included evergreen trees and shrubs of the tropics, subtropics, and the Mediterranean; temperate evergreen conifers like the Scotch pine (Pinus sylvestris); deciduous trees and shrubs of the temperate regions, e.g. European beech (Fagus sylvatica) as well as weedy plants like the water hyacinth (Eichornia crassipes), lambsquarters (Chenopodium album), bindweed (Convolvolus arvensis), and wild oat (Avena fatua).

Examples of C4 species:

Those of economic importance such as corn or maize (*Zea mays*), sugarcane (*Saccharum officinarum*), sorghum (Sorghum bicolor), and millets, as well as the *Panicum virganum*, which has been utilized as a source of biofuel. Other examples are: serious weeds such as *Cyperus rotundus, Cynodon dactylon,* (*Echinocloa* spp.), *Eleusine indica, Sorghum halepense, Imperata cylindrica, Portulaca oleracea, Digitaria sanguinalis, Amaranthus* spp., *Paspalum conjugatum, Rottboellia exaltata,* and *Salsola kali.*

Table 2. Some characteristics and the general distribution of the three plant types grouped according to their mechanisms of photosynthesis in the Dark reactions

Attribute/ Characteristic	C3 Plants	C4 Plants	CAM Plants
Distribution in the plant kingdom (% of plant species)	~85%	~3%, all angiosperms including most troublesome weeds; mostly monocots (C4 grasses and sedges about 79% of all C4 plants)	~8%, mostly succulent plants but not all succulents are CAM plants
Type of photosynthesis	C3 photosynthesis	C4 photosynthesis	CAM photosynthesis
CO2 fixation pathway	via C3 cycle only	via C3 and C4 cycles, spatially (C4 in the mesophyll cell then C3 in the bundle sheath cell)	via C3 and C4 cycles, both spatially (in different parts of same cell) and temporally (C4 at night, C3 at day time)
Leaf anatomy	Large air spaces bordered by loosely arranged spongy mesophyll cells; mesophyll cells but not bundle sheath cells (BSC) contain chloroplasts	Generally thinner leaves, closer arrangement of vascular bundles, smaller air spaces than C3; veins surrounded by thick-walled BSC further surrounded by thin-walled mesophyll cells (wreath-like arrangement of BSC is called Kranz anatomy); mesophyll cells and BSC contain chloroplasts, those of the BSC much larger	Thick and fleshy leaves, mesophyll cells having large, water-filled vacuoles
Stomatal movement	Stomata open at daytime, close at night	Stomata open at daytime, close at night	Inverted stomatal cycle (open at night, close in the day)
Typical Environmental / Geographical adaptation (where most common)	Temperate	Tropical or semi-tropical, high light intensity, high temperature, drought conditions	Desert or arid (xeric) habitat

Examples of CAM Plants

CAM plants are those that photosynthesize through Crassulacean Acid Metabolism or CAM photosynthesis. They are exceptionally succulent plants (fleshy plants having a low surface-to-volume ratio) although not all succulents belong to this plant type. Many halophytes (plants adapted to salty soils) are succulents but are not CAM. These plants are adapted to dry, desert habitats. They comprise up to about 20,000 species in about 40 families, equivalent to some 8% of all land plants. They are found in the families Isoetaceae (lycophytes), Polypodiaceae (ferns), Vittariaceae (ferns), Zamiaceae (cycads), and Welwitschiaceae (Gnetales), as well as in numerous families in the angiosperms (Agavaceae).

Specific examples of CAM plants are: *Crassula argentea*, Aeonium, Echeveria, Kalanchoe, and Sedum of the family Crassulaceae, *Ananas comosus*, *Tillandsia usneoides*, cacti, orchids, Agave, and wax plant (*Hoya carnosa*, family Apocynaceae).

The presence of both C3 and C4 species can be desirable in a pasture as they can occupy different niches (e.g. C3 species are often more abundant in the shade of trees and on southerly aspects, while C4 species often dominate full-sun conditions and northerly aspects) and thereby provide greater groundcover across a range of conditions. It is common to find both C3 and C4 species in one enclosure. This has advantages in providing a broader spread of production throughout the year for both grazing enterprises and native animals.

The Carbon cycle

The carbon cycle is referred as the flow of carbon through different parts of the Earth system (including the air and the bodies of plants and animals). The carbon cycle is a natural and integral part of life on Earth. A single carbon atom stored in a blade of grass may shift to the body of an animal that eats the grass. When the animal dies, its body might rot, and the carbon atom could join with oxygen to form CO_2 in the air. From there it might be taken in by a tree in the process of photosynthesis and used as a building block in a branch or trunk, or absorbed by the ocean. And so on. Although the basic flows of the carbon cycle have not significantly changed, in the last century or so, humans have increased the amount of CO_2 in the air by taking carbon that has been locked up in the ground for millions of years (in the form of oil, coal and gas) and releasing it into the atmosphere by burning those fuels. There is strong evidence that this has led to global warming.

Types of grasses

By their nature they can be:

- Natural or native
- Artificial or cultivated. They are exotic introduced usually are much improved and with good performance.
- Naturalized. Are those that were introduced and currently are predominate in greater proportions, e.g. common Buffel grass (*Cenchrus ciliaris* L.).

For its use can be:

- Grazing. All those pastures of lading or lower size.
- Grazing and cutting. They are grasses that are not used in courts.
- Court. They are the high court.

By its chemical composition are divided into:

- Complete. Are those with balanced nutrients, containing all the elements necessary for animal growth.
- Incomplete. Are those with low nutrient content because forage does not meet the requirements of the animals.

Use and conservation of pastures

When choosing a green forage grass, you should examine its possible uses and the final yield will provide to livestock. In Table 2.2 are shown the losses associated with the various uses of pasture. Various degrees of human interference or improvement have been usually applied to natural grasslands and more particularly to areas where the hay is cut. Fire is a powerful tool for managing rangeland, especially for the control of woody species and the removal of aged fodder. The introduction of animals, domesticated or wild has a great effect on vegetation. Handling or grazing pressure and control lead to changes in the botanical composition without deliberate introduction of species. The cleaning of the bushes, fences, drainage, application of fertilizers and trace elements are intensive interventions that alter the natural vegetation of the pastures. The introduction of grasses and legumes, without much cultivation is in any way other modifications stage. Many good natural grasslands have been replaced by arable while, in some countries, marginal agricultural land has been reverted to natural grasslands as the declining crop productivity.

Tabla 1. 2. Yield of grasses according to their use

Lost/yied	Grazing	Shop Green forage	Hay	Silage	Stored
Loss during crop, %	50	2	25	5	10
Loss during storage, %	-	-	5	30	1
Loss during transport, %	-	15	15	15	1
Total loss of dry matter, %	50	17	45	50	13
Forage effectively offered to animal, %	50	83	55	50	87
Maximum yield of dry matter, %	60	60	92	92	92
Potential yield of dry matter, %	30	50	51	46	70

Obtained from: Skerman y Riveros (1992).

Grazing

Grazing is the most common form of use of forage grasses. Plants are chosen for their abundance, by supporting various defoliation periods during grazing, trampling resistance and response to fertilizer use and its palatability, availability and nutritional quality. However, several problems such as loss of material by trampling, fecal contamination, selective grazing and early maturity in relation to the number of animals that graze; thus increasing the proportion of stems that becomes less digestive the pastures. In intensive and semi-intensive farms, subdivision controls by permanent and temporary fences are required and they may be electrified. Grazing requires less labor and takes less time than other methods of feeding. The animal choose their diet both in quality and quantity and nutrients are returned to the soil through the feces.

Grass hay

Hay is the most common way to store the grass. It is used to satisfy the food supply throughout the year. Usually it is the most convenient storage. The aim is to preserve the maximum amount of dry matter and nutrients at the lowest cost. Hay should be prepared at the optimum time for maximum yields and have the digestible dry matter required to meet the nutritional needs of cattle. Ideally, when cut at the beginning of the flowering stage. If you cut before, the nutritional value is higher but performance decreases and the moisture content is too high, making it difficult to cure. If you cut after flowering, increases in performance will not offset the decline in palatability and nutritional value. The first cut of grass hay is usually the better quality than the following. To produce good quality hay is essential that the pastures dry quickly and this not exposed to too much to the sun. Harvesting hay with rakes can cause decreased quality hay for the loss of leaves. Rain can also cause loss of leaves and nutrient leaching. Under normal conditions, the hay causes a loss of up to 25% of nutrients from grass.

Biotechnology of grasses

The wide distribution and pasture development worldwide largely, is due to the morphology of seed that helps their dispersion, their high reproductive capacity and their high tolerance to different types of environmental restrictions; for example, many of the drought-tolerant plants are in the family Graminae. Almost all temperate grasses have a basic chromosome number seven (common in cereals such as wheat, barley, oats and rye), while in tropical grasses haploid genome is 8, 9 or 10 chromosomes.

Biotechnology involves a group of scientific and technological procedures and tools used to quickly and efficiently develop a wide variety of processes and products, through the manipulation of living organisms or parts of them. It involves a set of processes of molecular and cell biology, genetics, biochemistry and microbiology that allow manipulation of organisms to increase productivity, improve specific characteristics of living things, produce new varieties of organisms, reduce production risks and develop new products or bioprocesses for preparing these, in using for biological systems.

Biotechnology includes pasture plant micro-propagation (pathogen-free plants or massive multiplication of particular genotypes), genetic transformation, somatic hybridization, use of bio-fertilizers. Actually, regeneration systems have developed for about 70 forage, turf grasses, ornamentals, and for biofuel.

The generation of genetically modified plants has resulted in the following:

1. Increased tolerance to different types of environmental stress
2. Fruits lower allergenic properties, greater resistance to cold and better nutritional value
3. Flowers with modified colors
4. Plants used for soil remediation
5. Trees with better solubilization and fragmentation of lignin
6. Plants used of colored fibers
7. Plants used as bioreactors for producing proteins, carbohydrates and specific oils.

CHAPTER 3

Nutrition of grass carbohydrates

Introduction

As livestock production continues to improve through breeding, the need for high quality forage has also increased. The producers control the quality of selecting fodder harvest or grazing date. This should increase the emphasis on a better understanding of the effect of environmental factors on forage quality. Unfortunately, the mechanisms by which environmental factors influence the quality of forage are not well understood, particularly at the molecular level and our understanding is not enough to predict the influence of environmental factors. Usually, the temperature has great influence on forage quality, more than other environmental factors, and is, in this particular area, where more information is needed. The increase in temperature normally causes maturity, however the primary effects on digestibility may be through the effect of the relationship between leaves and stems. High temperatures promote the growth of the stem over the leaf growth. The digestibility of stems and leaves is low in forages warm weather due to the high concentrations of cell wall and low content of non-structural carbohydrates. An increase in temperature can have a positive effect on forage quality by raising the concentration of CP. Soil nutrients have only small effects on forage quality. N fertilization, usually increase CP levels of some non-leguminous fodder. Forage species with low concentrations of N, such as winter pastures can improve digestibility, because N fertilization can stimulate microbial activity in the rumen. Additionally, the application of sulfur in deficient soils, this often stimulates digestibility. Foliar diseases probably have the most adverse effects on the quality of the forage plant. Pesticides in plants may reduce digestibility.

Forage

In general, forages are the vegetative parts of grasses containing a high proportion of fiber (more than 30% of neutral detergent fiber). They are required in the diet in a coarse physical form (particles larger than 1 or 2 mm in length). Forages usually occur: 1) at the farm, 2) directly grazed, and 3) harvested and preserved as silage or hay. Depending on the stage of growth, forages can contribute from almost 100% (in non-lactating animals) and not less than 30% (in cows in early lactation) of the dry matter in the diet. The general, characteristics of forages are:

1. The volume limits consumption of the ruminant. Too much forage in the diet may limit energy intake and milk production. However, bulky feeds are essential to stimulate rumination and maintain the health of the ruminant.
2. They can contain 30-90% of neutral detergent fiber (NDF). In general, the higher the fiber content, the lower the energy content of the forage.
3. Depending on the maturity, legumes may contain 15-23% crude protein (CP), grasses; however, contain 8-18% CP (according to the level of nitrogen fertilization) and crop residues (straw or stubble) may have only 3-4% of CP.

From a nutritional standpoint, forages may range from very good feeds (lush young grass, legumes at a vegetative stage) to very poor (straw and roughage).

Grasses and Legumes

High quality forage can make up two-thirds of the dry matter in the diet of ruminants, that consumes 2.5 to 3 % of their body weight (a cow of 600 kg can eat 15-18 kg of dry matter of a good quality forage). Cows can eat more than one legume than grasses at the same stage of maturity. However, good quality forages in good balanced diets, can provide much of the protein and energy needed for milk production.

The soil and climate conditions typically determine the most common types of forages in a region. Both grasses and legumes are widely known around the world. Grasses need nitrogen fertilizers and moisture conditions to grow well. However, legumes are more resistant to drought and require less N in the soil because they live associated with bacteria that can convert air N to soil N for fertilization.

The nutritive value of forages is highly influenced by the stage of growth when are harvested or grazed. Growth can be divided in three successive stages:

1. Vegetative stage
2. Flowering stage and
3. Seed formation stage

Usually, the feeding value of a forage is highest during vegetative growth stage and the lowest during seed formation stage. As maturity progresses, the concentration of CP, energy, calcium, phosphorus and digestible dry matter in the plant is reduced; while, increasing the concentration of NDF. When NDF is increasing, lignin content increases, making less available the carbohydrates to the rumen microbes. Thus, the energy value of the forage decreases. Therefore, when forages are produced for feeding cattle, they have to be harvested or grazed at an early stage. Corn and sorghum harvested for silage are exceptions because, despite the nutritive value of the vegetative parts of the plant (stem and leaves), during the seed formation, a high amount of digestible starch accumulates in the grains. The maximum yield of digestible dry matter of a forage crop is obtained:

1. In grasses, during the first part of maturity
2. In legumes, at the stage of mature button medium
3. In corn and sorghum, before the grains are fully completed

The nutritional value of a forage is reduced with advanced of maturity. The delay of the harvest after the optimum maturity may reduce the potential animal production of cattle consuming forage. However, several strategies are available to maintain the availability of forage that has good nutritive value:

1. Develop a grazing strategy that matches the number of animals in a pasture and the rate of grass growth.
2. Plant a mixture of grasses and legumes that have different rates of growth and maturity throughout the season.
3. Harvest at an early stage of maturity and preserve as hay or silage.
4. Feed lower quality forage to dry cows or the cows in late lactation, and the good quality forage to the cows in early lactation.

Crop residues and byproducts

The residues are the parts of plants that remain in the field after harvesting the main crop (e.g. corn roughage, cereal straw, sugar cane bagasse, peanut hay). The residues may be grazed, processed as dry feed, or made into silage. General characteristics of most residues are:

1. They are cheap and bulky foods
2. High in indigestible fiber because of its high lignin content. Although chemical treatments can improve its nutritional value
3. Low crude protein
4. Require adequate supplementation especially with protein and minerals
5. Require be chopped when harvested or before feeding
6. Can be included in the diets of lactating cows that have lower energy demands.

Concentrates

There is no clear definition of the concentrates, but can be described by their characteristics as food and its effects on rumen function. Usually, the concentrates can be referred as:

1. They are low in fiber and high in energy.
2. They can be high or low in protein. Cereal grains contain <12% CP, but oilseed meals (soybean, cotton, peanut) called protein foods can contain >50% of CP.
3. They have high palatability and are usually eaten rapidly. In contrast to forage, concentrates have low volume per unit of weight (high specific gravity).
4. Do not stimulate rumination.
5. They usually ferment faster than forages in the rumen. Thereby, increasing acidity (lower pH) in the rumen may interfere with normal fiber fermentation.
6. When they comprise more than 60-70% of the diet may lead to health problems.

Lactating ruminants have high requirements for energy and protein. Because cows can eat only a certain amount each day, forage alone may not supply the required amount of energy and protein. The purpose of adding concentrates to the diet of lactating cattle is to provide a source of energy and CP to supplement the forage and meet the requirements of the animal. Thus, concentrates are important feeds that allow for formulating diets that will maximize milk production. Generally, the maximum amount of concentrates that a cow can receive per day should not exceed 12 to 14 kg.

Types of carbohydrates

Carbohydrates are the most important source of energy and are the main precursors of fat and sugar (lactose) in milk. Microorganisms in the rumen allow the ruminant to obtain energy from fibrous carbohydrates (cellulose and hemicellulose) that are bound to the lignin in cell walls of the plant (Table 3.1). The fiber is bulky and is retained in the rumen because the cellulose and hemicellulose are fermented slowly. As the plants mature, the lignin content of the fiber increases and the degree of fermentation of cellulose and hemicellulose in the rumen is reduced. The presence of fiber in the diet is necessary to stimulate rumination. Rumination increases the breakdown and fermentation of fiber, stimulates contractions of the rumen and increases the flow of saliva to the rumen. Saliva contains sodium phosphates that help maintain the acidity (pH) of the rumen contents to a nearly neutral pH. Diets low in fiber and high in concentrates result in a low percentage of fat in the milk and contribute to digestive disorders, such as displaced abomasum and rumen acidosis.

Table 3.1 Carbohydrates contained in plants

Component	Function	Fruits	Seeds	Legumes	Grasses	Trees and shrubs
Soluble sugars	Nonstructural	17-77	0-1	2-16	5-15	5-15
Starch	Nonstructural	0-3	80	1-7	1-5	--
Pectin	Structural	5-17	0-1	5-10	1-2	6-12
Hemicellulose	Structural	2-7	7-15	3-10	15-40	8-12
Cellulose	Structural	3-17	2-5	7-35	20-40	12-30

Obtained from: Robbins (2001).

Nonstructural carbohydrates (starch and sugars) are fermented rapidly and completely in the rumen. The content of nonstructural carbohydrates increases the energy density of the diet by improving energy supply and increasing microbial protein produced in the rumen. However, the nonstructural carbohydrates not stimulate rumination fermentation or saliva production and when are excess may inhibit fermentation of fiber. Therefore, the balance between structural and nonstructural carbohydrates is important in ruminant feed for efficient production.

Glucose synthesis in liver

All propionic acid produced in the rumen is converted to glucose in the liver. The liver uses amino acids for glucose synthesis. This is an important process because usually the glucose can be absorbed from the digestive tract and the liver produces all the sugars found in milk. The exception is when the cow is being fed large amounts of concentrates rich in starch or a source of resistant starch ruminal fermentation. The glucose formed by the digestion in the intestine is absorbed and transported to the liver where it contributes to glucose supply of the cow. Lactose is an alternative source of glucose for the liver. Lactose is found in well-preserved silage, but lactose production in the rumen occurs when there is excess starch in the diet. This is undesirable because the acid in rumen environment, fiber fermentation stops and, in extreme cases, the animal stops eating.

Lactose and fat synthesis in the liver

During lactation, the mammary gland has high priority for the use of glucose. Glucose is used primarily for the formation of lactose (milk sugar). The amount of lactose synthesized in the mammary gland is closely related with the amount of milk produced every day. The concentration of lactose in milk is relatively constant and water is added to the amount of lactose produced by the secretory cells to a concentration of approximately 4.5% lactose. Therefore, milk production in dairy cows is highly influenced by the amount of glucose derived from the propionic acid produced in the rumen. Furthermore, glucose is converted to glycerol that is used for the synthesis of milk fat. Volatile fatty acids (VFA) acetic acid and β-hydroxybutyric acid are used for the formation of fatty acids of milk fat. The mammary gland synthesizes saturated fatty acids containing from 4 to 16 carbon atoms (short chain fatty acids). Almost half of milk fat is synthesized in the mammary gland. The other half that is rich in unsaturated fatty acids containing 16 to 22 carbon atoms (long chain fatty acids) are derived from dietary lipids. The energy required for the synthesis of fat and lactose is obtained from the combustion of ketone bodies; however, acetic acid and glucose can also be used as energy sources for the cells of many tissues.

Carbohydrates and grass quality

Carbohydrates are the main reservoir of photosynthetic energy from plants. The nutritional characteristics of carbohydrates for animal feeds are variable and depend on their sugar components and their bounds. However, the variety of sugars and bounds in plants is much wide than animal tissues. Carbohydrates of plants contain many sugars and uncommon links than animal systems. The nutritional availability depends on the capacity to break the glycosidic linkages in the carbohydrates of plants and between carbohydrates and other substances. They form the bulk of the food supply for animals, and is the most abundant class of compounds found in plants. Play important roles such as 1) intermediary metabolism, 2) energy transfer, 3) storage and 4) plant structure. Photosynthetic energy is set to carbohydrates via the Calvin cycle and serve as initial substrates for intermediate pathways in almost all plants. The energy is transported within the plants as the disaccharide sucrose, and stored in polymers such as starch and fructans.

Carbohydrates also constitute most of the cell wall of plants and as such play an important role in the structural integrity of individual cells, tissues and organs. They are extremely important from a nutritional perspective, and are the main source of energy in the diet of a ruminant. In ruminants most of the carbohydrate digestion occurs within the rumen (over 90%), although under certain circumstances such as high passage rates, a significant amount of carbohydrate digestion can occur in small intestine and large intestine. Sugars are rapidly fermented in the rumen to give VFA that are absorbed into the blood through the rumen wall. The polysaccharides must be degraded into simple sugars before being used. Nonstructural polysaccharides such as starch and fructans are rapidly and completely degraded in the rumen, while the degradability of structural polysaccharides (cellulose and hemicellulose) varies considerably.

In general, the degradability of cellulose in forages ranges from 25 to 90%, while the digestibility of hemicellulose varies from 45 to 90%. The degradation of β-glucans is intermediate to cellulose. The ability to degrade and utilize structural carbohydrates gives to ruminants a unique ecological niche. Besides being an important source of energy in the diet of ruminants, carbohydrates have other nutritional roles as components of dietary fiber. Structural carbohydrates are important for normal rumen function. Fiber stimulates rumination and salivation as previously mentioned, and promotes the exchange of cations that are important in ruminal buffering capacity. The fiber is also involved in the regulation of voluntary intake.

Chemistry of carbohydrates of the forage

The terms fiber and cell wall of the plant are often misleading. These terms, however, are not synonymous and reflect different functional perspectives. Plants are unique among higher organisms, although they have rigid cell walls. The cell walls of plants can be considered a compound consisting of cellulose fibrils embedded within a matrix of lignin and hemicellulose polysaccharides (Figure 3.1). In addition, intact cell wall contains components such as water, organic solvents and phenolic that give unique properties to the structure (Table 3.2). The macromolecular composition of the cell walls of the cells varies considerably between organs, tissue and subcellular level. The primary cell wall is formed adjacent to the plasmalemma during cell elongation and consists almost entirely of polysaccharides. The secondary wall is formed during cellular differentiation inner wall and the primary composition varies greatly depending on the cell type. Individual cells stick to the middle lamella, which consists mainly of pectic substances that serve as an intercellular cementing agent.

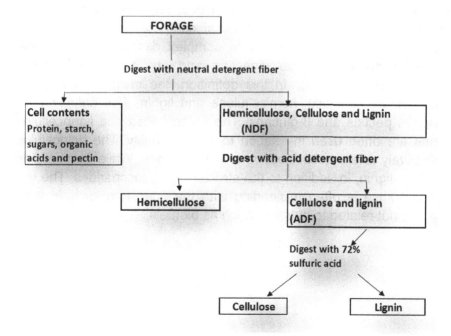

Figure 3.1. Schematic of detergent system of forage analysis

Table 3.2. Classification of forage fraction using the Van Soest Method

Faction	Components included	Ruminant
Cell contents	Sugars, starch, pectin	Complete
	Soluble carbohydrates	Complete
	Protein, Nonprotein	high
	Nitrogen, Lipids (fats)	high
	Water soluble vitamins	
	and minerals	
Cell wall (NDF)	Hemicellulose	Partial
	Cellulose	Partial
	Heat damaged protein	Indigestible
	Lignin	Indigestible
	Silica	Indigestible

The most obvious function of the cell wall (Figure 3.2) is its role in morphogenesis. The cell walls form the structural design of the architecture of the plant and provide mechanical and structural support to the plant organs. Also, the walls play important roles in water balance, ion exchange, cell recognition and protection of biotic stress. In contrast, the fiber is a nutritional entity which is defined by its biological properties and chemical composition. The concept of fiber, particularly fodder, refers to complex dietary nutrients, which are relatively resistant to digestion and are slowly and partially degraded by ruminants. In this definition, the main components of the fiber are cellulose, hemicellulose and lignin. This definition also includes pectins and β-glucans. The holocelulosa and lignocellulose terms are often used in relation to forage quality. The holocelulosa collectively refers to cellulose and hemicellulose. While lignocellulose includes lignin, in addition to the structural polysaccharides. The term lignocellulose is often misleading with the fiber, especially in areas that are not related to nutrition, such as biofuels.

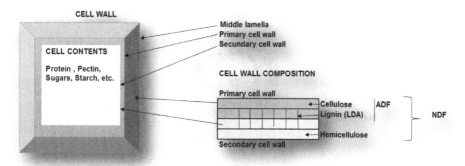

Figure 3.2. Schematic representation of plant cell wall composition
(ADF = acid detergent fiber; NDF = neutral detergent fiber)

Biosynthesis of carbohydrates

Carbohydrates are produced by photosynthetic carbon fixation process (Figure 3.3). The formation of individual types of sugars typically occurs through the action of the enzymes epimerases, isomerases, oxidoreductases and/or decarboxylase, of activated monosaccharides leaving Calvin cycle or from the breakdown of storage carbohydrates. The biosynthesis of oligosaccharides and polysaccharides requires activated sugars, in the form of nucleoside diphosphate monosaccharides. The predominant pattern of interconversions of glucose is derived directly from photosynthetic activity or starch degradation. There are a few alternate routes such as the conversion of inositol or glucuronic acid degradation pathways that can claim the galactose and galacturonic acid through direct phosphorylation.

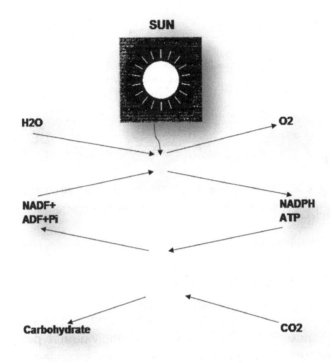

Figure 3.3. Reactions of carbon fixation for the biosynthesis of carbohydrates

Variation in composition of structural carbohydrates

There is considerable variation between species of plants with respect to the concentration and composition of structural carbohydrates. The cellulose concentration is typically higher in the walls of legumes than in grasses. This reflects a much lower concentration of hemicellulose in legumes compared to grasses. The concentration of cellulose often appears similar between grasses and legumes. The warm weather perennial grasses contain structural carbohydrates in greater proportion than temperate grasses.

The hydrolysis of forage hemicellulose produces the neutral monosaccharides: glucose, xylose, arabinose, mannose, galactose, rhamnose, fructose, and the acids uronic, galacturonic, glucuronic and 4-0-metilglucorónic. The relative proportions of each monosaccharide vary between species, reflecting differences in the structure of polysaccharides. Xylose and arabinose are produced from most neutral sugars isolated of hemicelluloses from grasses and legumes. Comparative studies were conducted in neutral structural carbohydrates isolated from stems of legumes and grasses. Glucose (predominantly from native cellulose) and xylose comprised 67% and 20% from the cell wall of legumes and 63% and 30% in grasses, respectively. It was recognized that the degradation of cell wall polysaccharides is much more affected by interactions between cell wall polymers, than for the individual properties of the polymers. Cellulose is degraded in the rumen by a complex of anaerobic microorganisms including bacteria, protozoa and fungi.

The cellulolytic bacteria from which *Ruminococcus flavefaciers*, *Fibrobacter succinogenes* and *R. albus* are the most important, and are responsible for cellulose digestion that occurs in the rumen (Figure 3.5). Although ciliated protozoa and fungi have been identified in the rumen that have cellulolytic activity, their contribution is relatively minor to the degradation of cellulose. Cellulolytic bacteria adhere to the surface of the cell wall, placing enzymes in close proximity to the substrate. The cellulolysis is accomplished by the action of several extracellular enzymes that bind to the surface of the body or are secreted into the surrounding medium. However, three basic enzyme activities are involved: 1) endo-β-1, 4-glucanase that breaks the polysaccharide random into oligosaccharides, 2) exo-β-1, 4-glucanase which attacks the no reducing end of oligosaccharides, giving cellobiose, and 3) cellobiose β-1, 4-glucosidase that hydrolyses cellobiose to glucose. The amount by which the native cellulose is used by rumen microorganisms is limited because it is association with lignin and other cell wall constituents.

There are, however, inherent factors that can limit the rate at which the cellulose is digested. The crystallinity of the cellulose has been suggested as a factor in reducing the accessibility of the cellulose to enzymatic attack. Cellulose degradation has been shown to be inversely proportional to the degree of crystallinity for purified substrates. However, actually, there is little evidence that the crystallinity is a limiting factor in the rate of degradation of native cellulose by rumen microbes.

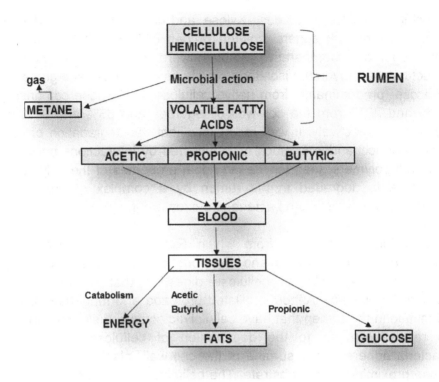

Figure 3.5. Ruminal digestion and absorption of carbohydrates

Hemicellulose degradation in the rumen occurs in a manner analogous to that of cellulose, but involves a broader arrangement of enzyme activities. The same cellulolytic bacteria listed above are responsible for most cellulose degradation in the rumen and are for the major hemicellulosic bacteria. In addition, *Butirivibrio fibrisolvens* that has a relatively minor role in the degradation of cellulose has a proportionally greater role in the degradation of xylans.

Some fungi and rumen protozoa have also hemicellulolytic activity, but its activity in the degradation of hemicellulose is relatively minor compared to ruminal bacteria. Isolated hemicelluloses are general, completely digested by rumen microorganisms; degradation of hemicellulose occurs through activities endo and exo glycanases that depolymerize and solubilize the major polysaccharide chains. Substitutes groups and side chains are removed from the hemicellulose and subsequently degraded by several glucosidases.

Cell wall (NDF) content in cultivated grasses

Seasonal NDF, cellulose and hemicellulose content of in cultivated grasses grown in different counties of the state of Nuevo Leon, Mexico are listed in Table 3.3. The NDF average in grasses was 76% with values ranging from 74 to 78%. Cellulose was slightly higher (34% annual mean) than hemicellulose (31%) in most grasses. However, in *Cynodon dactylon* the hemicellulose (34%) was higher than cellulose (32%) content. Due to small differences between grasses on the cellulose and hemicellulose contents, it may indicate that are digested by bacteria in the rumen in similar proportions. Differences in digestion might be due to genetic variations between grasses.

The NDF content of the six genotypes of the grass *Cenchrus ciliaris* was similar between collections (Table 3.4). However, cellulose doubled, in all cuts, the hemicellulose content. The same tendency as the NDF and its constituents (cellulose and hemicellulose), was observed in the 84 new genotypes of *C. ciliaris* (Table 3.5). Cellulose is degraded in the rumen by a complex anaerobic microorganisms among which are included bacteria, protozoa and fungi. *Ruminococcus flavefaciens* cellulolytic bacteria, *R. albus* and *Fibrobacter succinogens* are most important and are responsible for most of the digestion of cellulose in the rumen. Digestion of hemicellulose in the rumen occurs in the same manner as cellulose. The same bacteria mentioned above are responsible for digestion of hemicellulose, although it also includes *Butyrivibrio fibrisolvens*, but has much less effect than the other ones. Some fungi and protozoa also digest the cellulose and hemicellulose in the rumen, but to a much lesser extent than bacteria.

Cultivated grasses such as C. *ciliaris*, C. *dactylon*, D. *annulatum* and P. *coloratum* (Table 3.6) growing in the county of Linares, Nuevo Leon, Mexico, contained more NDF in stems to the leaves; additionally, cellulose and hemicellulose were largely higher in the stems than in the leaves. During wet seasons (spring and fall) grasses had less NDF content compared to the dry seasons (winter and summer).

Table 3.3. Seasonal content of neutral detergent fiber (NDF), cellulose y hemicellulose (Hemicel, % dry matter) in cultivated grasses collected in different counties of the state of Nuevo Leon, Mexico

Grasses	Place and date of collection	Concept	Winter	Spring	Summer	Autumn	Annual mean
Cenchrus ciliaris	Marin, N.L., México (1994)	NDF	78	74	76	74	76
		Cellulose	36	35	35	34	35
		Hemicel	32	30	31	33	32
Cenchrus ciliaris	Teran, N.L., México (2001-02)	NDF	81	73	86	68	77
		Cellulose	35	31	36	32	35
		Hemicel	32	30	30	30	32
Cenchrus ciliaris	Linares, N.L., México (1998-99)	NDF	72	72	70	71	71
		Cellulose	35	32	33	32	33
		Hemicel	33	30	36	26	31
Cynodon dactylon	Linares, N.L., México (1998-99)	NDF	81	72	83	78	79
		Cellulose	33	33	30	34	32
		Hemicel	32	27	44	32	34
Cynodon dactylon	Marin, N.L., México (1994)	NDF	69	77	82	76	76
		Cellulose	40	37	39	36	38
		Hemicel	25	26	21	27	25
Cynodon dactylon II	Marin, N.L., México (1994)	NDF	69	81	81	76	77
		Celulose	29	28	30	25	28
		Hemicel	29	37	40	36	35
Dichanthium annulatum	Marin, N.L., México (1994)	NDF	78	77	78	79	78
		Cellulose	29	34	38	33	33
		Hemicel	33	37	34	34	35
Dichanthium annulatum	Linares, N.L., México (1998-99)	NDF	78	77	83	77	79
		Cellulose	38	38	37	39	38
		Hemicel	22	26	32	24	26
Panicum coloratum	Linares, N.L., México (1998-99)	NDF	80	77	82	76	79
		Cellulose	33	34	33	35	34
		Hemicel	35	33	40	32	35
Rhynchelytrum repens	Teran, N.L., México (2001-02)	NDF	73	69	74	73	72
		Cellulose	28	27	29	28	28
		Hemicel	24	29	28	26	27
Mean		NDF	75	76	78	76	76
		Cellulose	34	33	35	33	34
		Hemicel	29	32	32	30	31

Obtained from: Ramírez et al. (2002); Ramírez et al. (2003); Ramírez (2003); Ramírez et al. (2004); Ramírez et al. (2005);

Table 3.4. Neutral detergent fiber (NDF), cellulose y hemicellulose (% dry matter) of the hybrid buffel Nueces and five new genotypes of *Cenchrus ciliaris* collected in different dates in the county of Teran of the state of Nuevo Leon, México

Genotypes	Concept	Dates of collections			
		Aug 1999	Nov 1999	Nov 2000	Jun 2000 (Fertilized)
Cenchrus ciliaris Nueces	NDF	74	71	72	72
	Cellulose	38	35	38	40
	Hemicellulose	25	27	24	24
Cenchrus ciliaris 307622	NDF	71	71	70	71
	Cellulose	38	35	33	43
	Hemicellulose	23	24	18	19
Cenchrus ciliaris 409252	NDF	73	69	70	70
	Cellulose	40	33	40	40
	Hemicellulose	24	26	21	18
Cenchrus ciliaris 409375	NDF	73	69	70	73
	Cellulose	39	35	38	40
	Hemicellulose	23	24	19	23
Cenchrus ciliaris 409460	NDF	72	69	70	74
	Cellulose	40	33	36	37
	Hemicellulose	23	22	22	24
Cenchrus ciliaris 443	NDF	71	66	66	69
	Cellulose	38	36	40	40
	Hemicellulose	24	20	15	20
Means	NDF	72	69	70	71
	Cellulose	39	35	38	40
	Hemicellulose	24	24	20	21

Obtained from: Garcia-Dessommes *et al.* (2003ab)

Cell wall (NDF) content in native grasses

Native grasses that grow in Marin county of the state of Nuevo Leon, Mexico contained about 10% more NDF (Table 3.7) than native grasses that grow in Teran county of the state of Nuevo Leon, Mexico (Table 3.8). The differences in NDF content may be due to differences in sites and dates of harvested. During the year of the study, in Marin county, the rainfall was 516 mm and in Linares was 613 mm. Thus, lower cell wall content in the grasses growing in Linares was because there was a greater precipitation. The cellulose content was similar to hemicellulose content in all native grasses (Tables 3.7 and 3.8). However, the grass *Cenchrus incertus* and *Chloris ciliata* had an intermediate NDF content when compared with other native grasses that grow in northeastern Mexico.

Table 3.5. Neutral detergent fiber content (NDF), cellulose (Cel) y hemicellulose (Hemi; % dry matter) in 84 new genotypes of *Cenchrus ciliaris* collected in the county of Teran of the state of Nuevo Leon, México in November 2000.

Genotypes	NDF	Cel	Hemi	Genotypes	NDF	Cel	Hemi	Genotypes	NDF	Cel	Hemi	Genotypes	NDF	Cel	Hemi
202513	73	38	29	409185	72	40	24	409258	67	39	21	409449	73	40	25
253261	72	38	27	409197	70	43	25	409263	70	40	27	409459	67	41	15
307622	69	33	18	409200	79	39	29	409264	73	41	27	409460	72	41	24
364428	69	38	22	409219	68	39	24	409266	74	41	31	409465	75	42	27
364439	76	42	23	409220	71	42	26	409270	73	41	29	409466	76	41	28
364445	73	40	22	409222	76	41	31	409278	71	43	25	409472	72	41	28
365654	72	41	23	409223	75	41	27	409280	70	42	26	409480	76	41	26
365702	72	41	27	409225	77	42	29	409300	75	41	23	409529	76	42	30
365704	71	42	21	409227	75	39	29	409342	70	40	25	409691	74	39	29
365713	75	40	27	409228	71	41	26	409359	73	43	23	409711	76	39	29
365728	76	41	25	409229	68	38	27	409363	68	40	25	414447	76	42	27
365731	77	41	27	409230	71	40	23	409369	70	41	23	414451	75	41	26
409142	75	40	27	409232	70	41	26	409373	69	37	19	414454	75	39	28
409151	72	42	26	409234	70	40	24	409375	72	41	24	414460	75	41	23
409154	70	41	19	409235	70	41	27	409377	71	40	24	414467	75	42	25
409155	71	42	26	409238	68	38	26	409381	71	40	27	414499	78	42	26
409157	71	42	23	409240	70	40	24	409391	75	40	24	414511	76	42	26
409162	69	41	22	409242	76	44	28	409400	75	40	24	414512	74	42	25
409164	74	40	22	409242	70	40	21	409410	74	40	25	414520	76	41	26
409165	69	39	22	409252	75	41	28	409424	70	41	21	414532	71	36	22
409168	70	40	24	409254	73	41	24	409448	76	41	30	443	72	38	24

Obtained from: Rodríguez-Morales (2003)

The environment and plant quality

There is no factor that affects the quality of the forage as the maturity of the plant; however, environmental changes the impact of maturity. The environment includes biotic and abiotic factors that influence the growth and development of plants. Cumulative effects are integrated through physiological processes and reflected in: 1) the growth rate of passage, 2) growth rate, 3) production and 4) quality of forage. Variations between seasons and changes in the environment related to the geographical location affect the quality of the forage, even when forages are harvested at similar morphological stages. This makes it difficult to predict the quality of forage, which affects the behavior of animals consuming the forage.

Table 3.6. Neutral detergent fiber content (NDF), cellulose y hemicellulose (Hemi; % dry matter) in leaves and stems in cultivated grasses and collected at different counties and dates of the state of Nuevo Leon, México

Grasses	Places and dates of collection	Parts	Concept	Seasons				Annual
				Winter	Summer	Summer	Fall	mean
Cenchrus ciliaris	Linares, N.L., Mexico (1998-99)	Leaves	NDF	78	62	83	63	71
			Cellulose	35	28	26	29	30
			Hemi	29	25	40	25	30
		Stems	NDF	88	75	88	74	81
			Cellulose	40	35	36	35	36
			Hemi	33	28	36	27	31
Cynodon dactylon	Linares, N.L., Mexico (1998-99)	Leaves	NDF	77	80	82	76	79
			Cellulose	32	31	27	31	30
			Hemi	34	38	47	34	38
		Stems	NDF	82	73	84	80	80
			Cellulose	44	35	31	36	36
			Hemi	31	26	44	32	33
Dichanthium annulatum	Linares, N.L., Mexico (1998-99)	Leaves	NDF	76	73	80	74	76
			Cellulose	36	37	34	38	36
			Hemi	23	24	35	25	27
		Stems	NDF	84	82	84	83	83
			Cellulose	43	43	37	43	41
			Hemi	22	24	35	24	26
Panicum coloratum	Linares, N.L., Mexico (1998-99)	Leaves	NDF	80	72	78	68	75
			Cellulose	31	26	28	27	28
			Hemi	38	38	41	33	38
		Stems	NDF	85	81	86	83	84
			Cellulose	37	37	40	38	38
			Hemi	35	32	36	33	34

Obtained from: Ramirez *et al.* (2001ab); Foroughbackhch *et al.* (2001); Ramirez *et al.* (2003); Ramirez (2003); Ramirez *et al.* (2005);

Plants rarely grow in their ideal environment, instead experience fluctuations in the environment and stress that: 1) changes its morphology and growth rate, 2) limits its production and 3) alter its quality. Stress is caused when some environmental factor is not ideal for plant growth and development. This can be caused by many factors but those that must be considered are 1) temperature, 2) water deficit, 3) solar radiation, 4) nutrient deficiencies and 5) pests. The cell wall of plants provides the first line of defense against most stresses. Cells develop secondary wall lignification, which is an important aspect of protection. Lignification also restricts the availability of nutrients in the cell wall for animals that consume them. The cell walls vary in digestibility, they are available only in part; however, the cell contents are completely digestible.

Table 3.7. Seasonal content of neutral detergent fiber (NDF), cellulose y hemicellulose (% dry matter) in native grasses collected in Teran county of the state of Nuevo Leon, México in 2001 and 2002

Grasses	Concept	Seasons and years of collection				Annual
		Winter 2002	Spring 2002	Summer 2002	Fall 2001	mean
Bouteloua curtipendula	NDF	72	72	73	79	74
	Cellulose	28	28	29	31	29
	Hemicellulose	27	27	29	29	28
Bouteloua trifida	NDF	75	70	76	76	74
	Cellulose	30	28	30	30	30
	Hemicellulose	27	26	30	29	28
Brachiaria fasciculata	NDF	72	63	62	60	65
	Cellulose	27	23	23	22	24
	Hemicellulose	25	22	25	22	24
Digitaria insulares	NDF	70	67	65	71	68
	Cellulose	33	32	31	33	32
	Hemicellulose	34	31	29	31	31
Chloris ciliata	NDF	70	70	74	75	72
	Cellulose	26	26	27	28	27
	Hemicellulose	26	27	26	26	26
Leptochloa filiformis	NDF	75	67	73	67	70
	Cellulose	29	25	28	25	27
	Hemicellulose	31	21	29	25	26
Panicum hallii	NDF	74	67	68	71	70
	Cellulose	31	28	28	30	30
	Hemicellulose	31	26	29	31	29

Panicum obtusum	NDF	74	65	66	66	65
	Cellulose	28	25	25	25	26
	Hemicellulose	28	23	24	24	24
Panicum unispicatum	NDF	70	64	69	67	68
	Cellulose	29	27	29	28	28
	Hemicellulose	32	26	28	26	28
Setaria grisebachii	NDF	73	71	61	73	69
	Cellulose	28	28	23	28	26
	Hemicellulose	26	27	21	26	26
Setaria macrostachya	NDF	72	68	63	73	69
	Cellulose	28	27	25	28	27
	Hemicellulose	28	27	23	25	26
Tridens eragrostoides	NDF	71	74	72	76	73
	Cellulose	30	31	30	32	30
	Hemicellulose	29	31	31	31	30
Tridens muticus	NDF	76	72	73	78	75
	Cellulose	28	27	27	29	28
	Hemicellulose	31	28	28	28	27
Means	NDF	73	68	69	72	70
	Cellulose	31	29	29	30	30
	Hemicellulose	30	28	29	28	29

Obtained from: Cobio-Nagao (2004)

The stress caused by the environment has a greater effect on the production of forage than in the digestibility or other factors related to quality. The environment of the plant often demonstrates its great influence not only in forage quality by altering the relations between the stems and leaves, but also cause other morphological changes in chemical composition of plant parts. Changes in the morphology of plants can alter the availability of forage, especially influencing consumption affecting grazing animals potential bite size. The height cover of the vegetation is the most important variable affecting grass bite size and altering the ratio stem:leaf. Environmental influence maturation rates and the amount of dead material. The animals generally selected young green tissues rather than the stems and dead leaves plant tissues. Many stresses reduce plant growth and development, the result is that the quantity of forage remains at very low levels.

Table 3.8. Neutral detergent fiber content (NDF), cellulose y hemicellulose (% dry matter) in native grasses collected in Marin county f the state of Nuevo Leon, México in 1994

Native grasses	Concept	Seasons				Annual
		Winter	Spring	Summer	Fall	mean
Aristida longiseta	NDF	87	85	87	88	87
	Cellulose	37	37	37	37	37
	Hemicellulose	37	33	37	37	37
Bouteloua gracilis	NDF	90	81	82	77	82
	Cellulose	33	38	32	33	34
	Hemicellulose	43	30	32	29	34
Cenchrus incertus	NDF	80	74	80	75	77
	Cellulose	29	28	35	30	31
	Hemicellulose	36	.26	34	33	32
Hilaria belangeri	NDF	83	75	82	75	79
	Cellulose	39	28	38	32	32
	Hemicellulose	32	31	31	30	31
Panicum hallii	NDF	76	69	73	68	72
	Cellulose	32	30	29	31	31
	Hemicellulose	31	29	32	26	29
Setaria macrostachya	NDF	80	79	86	74	80
	Cellulose	37	35	42	33	36
	Hemicellulose	32	35	35	31	33
Means	NDF	83	77	82	76	80
	Cellulose	35	33	36	33	34
	Hemicellulose	35	31	34	31	33

Obtained from: Ramirez et al. (2004)

Effect of temperature on nutritional quality of grasses

Because the nutritional value of a forage is regulated by the amount and availability of metabolic and anabolic products, including cell wall and cell contents, thus any factor that influences these products also affects the quality of forage. The temperature usually has a great influence on forage quality than other environmental factors found in plants. The temperature of the plant is the result of complex interactions between the plant and its environment and is influenced by the flux density of radiation, heat conduction, convection heat, latent heat and anatomical and morphological characteristics. Moreover, due to variations in coverage, particular aspects of plant parts and the result of differences in the radiation charge, the tissue temperature can vary widely at any time.

Temperature effects on plant development

In general, temperature and soil affect the quality of forage in certain species growing in certain regions. The temperature is the major determinant of the geographical adaptation of plant species. This is particularly manifested in the extreme temperatures encountered by ontogeny of plants. These extremes can cause plant death or severe weakness. Under field conditions, the high temperature stress often occurs along with the stress of water making it difficult to separate the two effects.

Within a region or area, the primary effects of temperature on the quality of grasses determine the rate of plant growth and influence relative proportion of leaves and stems. A side effect of the temperature difference is found in the tissue morphology of the leaves and stems. The temperature has a greater effect on the digestibility more than other environmental variables, the economic implications of changes in temperature should not be ignored.

Effect of temperature on chemical composition and digestibility

The negative effects of elevated temperature on the digestibility of the foliage of grasses have been subject to numerous studies over the last 30 years. In most of the plants that flower, environmental controls such as length of the day and temperature modulate the rate of development. The temperature cannot only increase the concentration of cell wall, but can also reduce cell wall digestibility. A decrease of 80 g kg^{-1} *in vitro* digestibility (IVDMD) in *Festuca arnudinacea* occurred when the temperature was increased from 15/10 °C to 25/20 °C. Results of several related experiments related with temperature and digestibility concluded the leaves of temperate grasses, during growth, showed an average decrease of 6.6 g kg^{-1} of IVDMD per each increase of one-Celsius degree in temperature. The decrease in IVDMD associated with high temperatures is most often attributed to high concentrations of the constituents of the cell wall, but little research has been conducted to test the causes involved in this phenomenon.

Winter and summer grasses

It has been reported that the geographical distribution of species C3 and C4 are determined by regional and seasonal temperatures. The C4 types are more numerous in the warmer and warmer seasons climates, C4 grasses have been recognized for their relatively low digestibility and high concentrations of structural polysaccharides compared with C3 grasses. In fact, in the grass, the difference in digestibility and composition of the cell wall are caused by the temperature. Within the *Festucoidae* and the tropical subfamily *Panicoidae*, the first exhibited the photosynthetic C3 pathway and the last had C4. Warm grasses also were high in cellulose and hemicellulose, rather than temperate grasses, this aspect has not been sufficient clarify whether the differences are truly taxonomic or are caused by different environmental conditions.

Using growth chambers with controlled temperature growth, it has been shown that an increase in temperature during growth of warm season grasses (*Brachiaria ruziziensis*) increased production of dry matter (DM), size, but also on the number of new leaves in leaf/stem ratio, and the concentration of organic nitrogen (N) in the DM. The authors also noted a positive relationship between temperature and concentration of crude fiber in both leaf and stem tissues and postulated that the temperature itself is the main factor contributing to the relatively poor quality grass in warm climates.

Effects of water on grass quality

Water is a crucial component of plant cells and is needed for all metabolic processes that depend on its presence. A suitable amount of water is required for maintenance of turgor pressure, the protection function and diffusion of the solutes into cells. Water provides oxygen during photosynthesis and the hydrogen used for the reduction of carbon dioxide. The amount of water varies with the cell type and the physiological status. The newly formed cells are necessarily composed of water, while fibrous consistency cells contain almost no water. On average, the concentration of water in grasses can be about 750 g kg^{-1}, depending on the species and environmental conditions, and decline as the maturity of the plant progresses.

Most of the water of grasses comes from the soil through the roots. The plants function as water pump, moving soil water into the atmosphere in response to differences in the potential of soil and air. About 1% of the water entering the growing plants is retained and most of it is lost through perspiration. Excess of water, for metabolic needs, usually is used for important functions in the movement of solutes from the roots to the leaves and stems and evaporative cooling of plant. Strong resistance to water movement through the plant normally occurs in the air spaces within the leaves. Stomata occupy about 1% of the leaf surfaces, but most water lost by living leaves, passes through these open organs. Some water is also lost through the cuticle.

General effects of water on grasses

The excess or deficiency of water can produce stress in grasses. Too much water that can result from waterlogged soils imposes stress because in waterlogged soils. The oxygen is lost by microorganisms and the root respiration, leaving the roots of grasses in an environment without oxygen. Even that anoxia can greatly reduce forage production; there is little information on how this affects the quality of grasses.

It is more common that most grasses are growing areas with dry soil than watery soils. In fact, the biggest concerns in the future will related to climate change and droughts. The water deficit stress is usually the greater physical limitation to grasses. When transpiration exceeds water absorption by the roots, the water deficit in the plant increases and stress can occur, which adversely affects many enzymatic reactions of most physiological processes. The water deficit causes stomatal closure reduces transpiration rates, and increases the temperature of the grass. Cell enlargement is particularly sensitive to water deficit.

Cell division appears to be less sensitive than cell enlargement. Turgor pressure plays an important role in cell enlargement, providing the necessary pressure to the cell wall expand. As the cell walls expand, decreased turgor pressure, which causes the water potential within the cell, decrease. This creates a difference between the inside and outside of the cell, which moves more water into the cells. Solutes should continue settling in growing cells.

The ability of pasture to maintain a constant positive or turgor under water potential decreases is an important adaptation process to water deficit. The most important physiological mechanisms allow plants maintain their turgor under water stress conditions in osmoregulation, which is the osmotic potential and can result from the condensation of the cells for water loss and increased solute in cells under conditions of water stress. Solutes that are in concentration include soluble sugars, organic amino acids. Under moderate to severe stress concentration of the amino acid proline, it increases more than other amino acids. Proline may serve as a store of N and an aid in drought tolerance acting as a solute in the osmoregulation. The photosynthetic rates are usually less affected by the drought that respiration rates and growth, resulting in an overall increase in concentration of nonstructural carbohydrates. The translocation of fotocinatos, however, is relatively sensitive to water deficit. The effect varies depending on the species level of stress and the state of development of the plant. Accumulation of nonstructural carbohydrates and stores of N can facilitate rapid regrowth after water stress is released.

Solar radiation

The first step in the use of solar energy is the conversion to chemical energy through photosynthesis. During this process, the energy flow starts in the ecosystem of the earth biosphere. The carbon used is fixed from atmospheric CO_2, representing around 0.03% of the total gas composition. Photosynthesis occurs when the green sheets are exposed to radiation in the visible part of the field (radiation is in a wavelength of 400 to 700 nm). The energy in these wavelengths represents about half the total solar radiation. Under ideal conditions up to 7% of the solar energy can be stored in photosynthetic products in fast-growing crops. However, in grasses throughout the growing season, the average is much lower (less than 1%).

Shading typically has a small effect on grass quality, compared with the morphology or production. It has been found that by imposing a 63% shade of 5 perennial grasses reduced by 43% the production and leaf 24% but, only reduced the concentration of NDF in 3%, the concentration of lignin in the cell wall in 4% and increased digestibility of forage in 5%. The N concentration is much more sensitive to shading than other quality characteristics. It has been found that 63% of the shadow the N concentration in grasses increased by 26%. The response was generally higher in the leaves than in the stems.

The cell wall components are deposited in the following order: hemicellulose, cellulose and lignin, although there are many overlap between these activities. The reduction in cell wall composition by shading is reflected in an increase in DM digestibility in some studies. It has been reported that forage digestibility was improved by 5% with heavy shading. Moreover, they found that DM digestibility of grasses developed under shade was higher than the grasses that grow under the influence of sunlight.

Interaction of environmental factors and plants

Among the climatic variables, light and temperature are the most important and then follows the water supply. This sequence becomes more noticeable in temperate climates. The growing season begins in spring, and then starts the slow growth of temperature and light faster, then the temperature reaching a maximum in the summer when the length of day begins to decline. Light, temperature and plant maturity have different effects on the composition of the plant, and these effects vary and interact differently in relation to the season of the year. The effect of irrigation, fertilization and predators should not be ignored. A summary of the main factors that influence and interact on the composition and nutritional value of forages is shown in Table 3.9.

Table 3.9. Environmental factors that influence and interact in composition and nutritional quality of forages

Parameter	Temperature	Light	Nitrogen	Water	Defoliation
Production	+	+	+	+	-
Soluble carbohydrates	-	+	-	-	+
Nitrates	-	-	+	NA	NA
Cell wall	+	-	±	+	-
Lignin	+	-	+	+	-
Digestion	-	+	±	-	+

Note:
+ = positive effect; - = negative association; ± = inconstant association; NA = not available data
Obtained from: Van Soest *et al.* (1978).

CHAPTER 4

Silica and lignin in Grass

Introduction

Silica and lignin are deposited on the cell walls of plants as part of the cell maturity. In grasses, silica and lignin are considered anti-nutritional components for its negative impact on the nutritional availability of plant fiber. They interfere with the digestion of cell wall polysaccharides by acting as a physical barrier to microbial enzymes. It has been confirmed that grasses receive silica to control biological paths, sustenance the plant and improve its lenience to salty conditions, heavy elements and toxic fungi. It has been demonstrated that silica stores mostly in the epidermis in the superior part of plants, in the center of sclerenchyma cells of the stele, the wall of endoderm, in the wall of vessel and in the sub epidermal tissues, epidermis and at the end of the roots. The mutual feature of deposition of silica in these cells is that it starts at the same period when the stop growing to produce and store lignin in the secondary wall of plants. It was decided that during this mechanism, lignin is penetrable to water and some soluble substances; however, it is no possible that it is penetrable to a great molecules such as monosilicic acid, consequential in the installation of silica. It has been also observed that silica was existent in the cell wall of rice beside with phenol-polysaccharide compounds or lignin-polysaccharide compounds. The silica may be related, as the element calcium, with lignin or phenol.

Silica

The silica in the soluble phase is in the form of orthosilicic acid (H_4SiO_4), resulting silicates that promote:

1) Creation of gradients of mineral elements from the soil to tissues of the plant
2) Affecting the accumulation and mobilization of carbohydrate reserves
3) Promoting the production of phytochemicals

For this to happen, it is ideal that orthosilicic acid concentration in the soil solution must be permanently higher than 70 ppm (approximately 35 kg ha^{-1} in the root zone). With this concentration, silica flow is endorsed to different plant tissues. One of these is the foliar and root epidermis, in which trichromes are developed. However, silica can form, together with the cutin, an indigestible wall on potentially degradable tissues (parenchyma and sclerenchyma collenchyma) and constitute a strong barrier, which prevents the necessary colonization of plant tissues by rumen bacteria, thereby decreasing their digestibility. In forages, for each unit of increase in silica content, decreased may occur up to 3% of the *in vitro* dry matter digestibility (IVDMD), mainly due to the decrease in the digestibility of the polysaccharides of the cell wall, and the inaccessibility to rumen microorganisms for colonization caused by silica. An equation was also established, that showed that the silica content is the best explanation of the decrease of IVDMD of the plant studied. It also constitutes a useful tool for establishing management strategies of the plant, which are most efficient for animal feed.

Lignin

The word comes from the Latin term lignum, meaning wood; and, those plants containing large amounts of lignin are called woody. The lignin is characterized as a complex aromatic (not carbohydrate) of which exist many structural polymers (lignins). It is convenient to use the term lignin in a collective sense to separate the lignin from the fiber fraction. After the polysaccharides, lignin is most abundant organic polymer in the plant world.

Lignin performs multiple functions that are essential to plant life. For example, it has an important role in internal transport of water, nutrients and metabolites. It provides rigidity to the cell wall and acts as a bridge between the cells of the wood, creating a material which is substantially resistant to impact, compression and twisting. Lignified tissues resist attack by microorganisms, preventing penetration of destructive enzymes in the cell wall.

Chemical structure of lignin

The lignin is a macromolecule with a high molecular weight, which results from the union of several phenylpropanoic acids and the alcohols (coumaryl, coniferyl and sinapyl). The randomized bounding of these radicals gives rise to a three-dimensional structure, an amorphous polymer that is characteristic of lignin. Lignin is a complex natural polymer with regard to its structure and heterogeneity. Therefore, it is not possible to describe a definite structure of lignin.

Physical properties of lignin

Lignins are polymers insoluble in acids and strong alkalis, which are not digested or absorbed and are not attached by the microflora of the large intestine. They can bind to bile acids and other organic compounds (e.g. cholesterol), delaying or decreasing absorption in the small intestine of such components. The lignification significantly affects the fiber digestibility. Lignin, increases in the cell wall of the plant during maturation. It is resistant to bacterial degradation and reduces the digestibility of structural polysaccharides.

Lignin is a polymer with no defined structure containing alcohols (hydroxycinamyl) and may contain phenolic compounds and phenolic acids. Lignin is often mentioned as limiting fiber digestion, and sometimes the protein. However, recent studies suggested that the lignin content itself would not be responsible for the decrease in fiber digestion, but lignin action would cause decrease of the access of hydrolytic enzymes to the fiber.

There are various methods for estimating lignin content in a food, being the best known, the digestion in concentrated sulfuric acid (72%). This method has been criticized because it overestimate the concentration of lignin in the forage, because the protein is also precipitated with lignin. The value of knowing the lignin concentration of a given food is it's apparently connection with digestibility or Indigestibility of that food. In this context, the effect of lignin on fiber digestibility appears greater in leguminous than in grasses. In general, as the growth stage of a given feed progresses, the concentration of lignin also increases.

Development of lignin

The inclusion of lignin in the cell wall begins in the middle lamella, within secondary wall. The effect of this is that the most recently deposited polysaccharides in the secondary wall are not lignified. The middle lamella and primary wall region is the most intensely lignified. This may explain why the rumen microbes degrade the cell wall of plants from the lumen outward, and because the middle lamella and primary wall region of lignified cells are never completely digested.

Relationship between cell wall and digestibility

Lignin is a component of the cell wall and is recognized as limiting the digestion of cell wall polysaccharides in the rumen. The effect of lignin on the digestibility of the grasses is assumed to have a direct influence on the digestibility of the wall rather than on the digestibility of the total organic matter of grasses. Lignin apparently has a negative effect on the digestibility of the polysaccharides of the cell wall because it protects the enzymatic hydrolysis of polysaccharides. The effect of lignin on fiber digestibility has been shown to be higher in grasses than in legumes. Several studies have demonstrated the negative correlation between the concentration of lignin and fiber digestibility or cell wall. The digestibility of mature cell wall is lower than that of immature cell walls; therefore, it is assumed that the lignin composition also affects the digestibility of the cell wall.

Lignin content in cultivated grasses

Grasses generally contain less lignin than legumes; however, as the maturity progresses, the lignin in legumes remains constant, but in the grasses increases. The annual average (7%) of lignin content in 13 cultivated grasses is shown in Table 4.1. Small variation in lignin content was observed among seasons. *Rhynchelytrum repens* had the highest content and *Cenchrus ciliaris* and *Cynodon dactylon* (Cross II) were lowest. Because lignin content limits ruminal digestion of grasses, thus, *R. repens* might have lower digestibility than other grass listed in Table 4.1.

In general, the mean (4.5%) of all genotypes of *Cenchrus ciliaris* (Tables 4.2 and 4.3) was lower than in most cultivated grasses shown in Table 4.1, except for *C. ciliaris* and *C. dactylon* (Cross II) that were lower. Low levels of lignin in genotypes may cause greater use of dry matter by rumen microorganisms causing increased the production of volatile fatty acids.

Table 4.1. Seasonal content of lignin (% dry matter) collected in different counties of the state of Nuevo Leon, México in different dates

Grasses	Places and dates of collection	Seasons				
		Winter	Spring	Summer	Fall	Annual mean
Cenchrus ciliaris, Common	Marin, N.L., Mexico (1994)	7	6	7	5	6
Cenchrus ciliaris, Common	Marin, N.L., Mexico (1998)	6	6	4	7	6
Cenchrus ciliaris, Common	Teran, N.L., Mexico (2001-02)	9	7	10	4	5
Cenchrus ciliaris, Common	Linares, N.L., Mexico (1998-99)	8	6	10	6	8
Cenchrus ciliaris, Nueces	Marin, N.L., Mexico (1998-99)	6	7	3	7	6
Cenchrus ciliaris, Llano	Marin, N.L., Mexico (1998-99)	8	8	5	10	8
Cynodon dactylon (Cross I)	Linares, N.L., Mexico (1998-99)	8	8	7	8	8
Cynodon dactylon (Cross I)	Marin, N.L., Mexico (1994)	7	7	6	10	7
Cynodon dactylon (Cross II)	Marin, N.L., Mexico (1994)	4	5	5	6	5
Dichanthium annulatum	Linares, N.L., Mexico (1998-99)	6	6	11	11	9
Dichanthium annulatum	Marin, N.L., Mexico (1994)	10	6	7	6	7
Panicum coloratum	Linares, N.L., Mexico (1998-99)	8	8	7	8	8
Rhynchelytrum repens	Teran, N.L., Mexico (2001-02)	10	8	10	10	12
Seasonal mean		8	7	7	8	7

Obtained from: Ramirez *et al.* (2002); Ramirez *et al.* (2003); Ramirez (2003); Ramirez *et al.* (2004); Ramirez *et al.* (2005);

Table 4.2. Lignin content (% dry matter) of the hybrid buffelgrass Nueces and five new genotypes of buffelgrass (*Cenchrus ciliaris*) collected in different dates in Teran county of the state of Nuevo Leon, México

Genotypes	Place of collection	Dates of collection			
		Aug.1999	Nov. 1999	Nov. 2000	Jun. 2000 (Fertilized)
Cenchrus ciliaris Nueces	Teran, N.L., Mexico	5	5	5	2
Cenchrus ciliaris 307622	Teran, N.L., Mexico	6	6	9	4
Cenchrus ciliaris 409252	Teran, N.L., Mexico	4	5	7	5
Cenchrus ciliaris 409375	Teran, N.L., Mexico	5	5	8	2
Cenchrus ciliaris 409460	Teran, N.L., Mexico	6	6	8	4
Cenchrus ciliaris 443	Teran, N.L., Mexico	5	5	7	4
Mean		5	5	7	4

Obtained from: Garcia-Dessommes *et al.* (2003ab)

The lignin content in stems was as twice as much of the leaves in cultivated grasses harvested in different counties of the state of Nuevo Leon, Mexico (Table 4.4). This could imply that the leaves are more digestible than stems because the latter contain more lignin. Lignification tends to be more extensive in structural tissues such as xylem and sclerenchyma. Plant organs containing high proportions of these tissues, such as stems, and are less digestible than those containing low concentrations. The proportion of woody tissues and these organs usually increases as the plant matures, thus, there is a negative relationship between digestibility and maturity. All these plant processes respond to environmental factors that can affect the amount and impact of woodiness. Temperatures, soil moisture, amount and quality of light and soil nutrient status may also have direct or indirect effects on lignification. Environmental stress (biotic or abiotic) that cause a reduction in stem ratio:leaf, usually decrease forage quality, because the leaves are more nutritious than stems.

Lignin content in native grasses

Native grasses (Tables 4.5 and 4.6) had more lignin than cultivated grasses (Tables 4.1, 4.2 and 4.3). This could be because the native grasses have higher stem:leaf ratio. With more stems increases the lignin content. The concentration and composition of lignin varies greatly among genera, species, and, to some extent, within species comprising forage plant communities. The grass *Aristida longiseta*, in all seasons, had the highest lignin content, but *Digitaria insularis* was lowest. In most native grasses very similar lignin content, among seasons, was observed. The most useful management practices for minimizing the negative effects of lignification are manipulation of the native grass community such that it contains species that are more desirable and harvest management to maintain plants in a vegetative stage of development. Grasses such as *Hilaria belangeri*, *Leptochloa filiformis* and *Panicum hallii*, had intermediate lignin values, thus, these grasses might be considered of good digestibility value. The goal of grass producers and managers is to develop a plant community with inherently high quality characteristics.

Table 4.3 Lignin content (% dry matter) of 84 new genotypes of buffelgrass (*Cenchrus ciliaris* L.) collected in the county of Teran of the state of Nuevo Leon, México in November 2000

Genotypes	Lignin	Genotypes	Lignin	Genotypes	Lignin	Genotypes	Lignin
202513	3	409185	4	409258	5	409449	2
253261	4	409197	4	409263	2	409459	5
307622	6	409200	8	409264	2	409460	2
364428	4	409219	3	409266	4	409465	5
364439	6	409220	2	409270	2	409466	3
364445	7	409222	2	409278	4	409472	5
365654	4	409223	2	409280	4	409480	4
365702	5	409225	3	409300	9	409529	4
365704	3	409227	4	409342	4	409691	4
365713	5	409228	5	409359	4	409711	4
365728	5	409229	2	409363	4	414447	2
365731	5	409230	3	409369	2	414451	5
409142	3	409232	4	409373	7	414454	2
409151	3	409234	2	409375	4	414460	6
409154	6	409235	4	409377	4	414467	4
409155	4	409238	4	409381	2	414499	6
409157	4	409240	2	409391	7	414511	5
409162	2	409242	2	409400	8	414512	4
409164	8	409242	5	409410	6	414520	3
409165	6	409252	5	409424	5	414532	8
409168	3	409254	2	409448	4	443	5
Mean of all genotypes							**4**

Obtained from: Morales-Rodríguez (2003).

There are a number of postharvest treatments that can be used to improve the digestibility of fiber in highly lignified forages. There are 4 basic strategies employed in postharvest treatments that have been developed to lessen the impact of lignin on fiber digestion. These include:

1) Alkaline hydrolysis
2) Enzymatic hydrolysis
3) Oxidation
4) Microbial treatments

Table 4.4. Lignin content (% dry matter) in leaves and stems of cultivated grasses collected in different counties and dates of the state of Nuevo Leon, México

Grasses	Places and dates of collection	Parts	Seasons				Annual
			Winter	Spring	Summer	Fall	mean
Cenchrus ciliaris Common	Linares, N.L., Mexico (1998-99)	Leaves	8	4	7	4	6
		Stems	13	9	13	9	11
Cenchrus ciliaris Common	Marin, N.L., Mexico (1998-99)	Leaves	4	3	3	5	4
		Stems	6	8	4	9	7
Cenchrus ciliaris Llano	Marin, N.L., Mexico (1998-99)	Leaves	4	6	2	4	4
		Stems	11	11	7	14	11
Cenchrus ciliaris Nueces	Marin, N.L., Mexico (1998-99)	Leaves	4	4	3	4	4
		Stems	10	11	5	11	9
Cynodon dactylon	Linares, N.L., Mexico (1998-99)	Leaves	7	6	4	6	6
		Stems	6	9	8	9	8
Dichanthium annulatum	Linares, N.L., Mexico (1998-99)	Leaves	7	5	4	6	6
		Stems	13	9	8	10	10
Panicum coloratum	Linares, N.L., Mexico (1998-99)	Leaves	6	5	5	5	5
		Stems	11	11	10	11	11

Obtained from: Ramirez *et al.* (2001ab); Foroughbackhch *et al.* (2001); Ramirez *et al.* (2003); Ramirez (2003); Ramirez *et al.* (2005).

Of these, alkaline hydrolysis is by far the most common and practical. Alkaline treatments have been demonstrated to improve the digestibility of grasses, but not legumes. The reagents most commonly used for this purpose are ammonia and various hydroxides. Ammoniation of forages can be accomplished using anhydrous ammonia applied in gaseous form or by incorporating urea with the forage. Ammoniation increases fiber digestibility and crude protein concentration whereas treatment with sodium hydroxide only increases fiber digestion, however, sodium hydroxide treatment generally results in greater increases in fiber digestibility than ammoniation. Of the alkaline treatments used to improve forage digestibility, ammoniation is the easiest to use. In all cases, chemical treatments used to enhance fiber digestion result in greater improvements when applied to poor quality roughages such as mature grasses and cereal crop residues than when applied to higher quality forages.

Table 4.5. Seasonal lignin content (% dry matter) of native grasses collected in the county of Marin of the state of Nuevo Leon, México 1994

| Native grasses | Seasons and year of collection | | | | Annual |
	Winter	Spring	Summer	Fall	mean
Aristida longiseta	10	10	9	10	10
Bouteloua gracilis	7	7	9	8	8
Cenchrus incertus	8	8	6	6	7
Hilaria belangeri	5	5	7	8	6
Panicum hallii	8	6	5	8	7
Setaria macrostachya	5	6	8	9	7
Seasonal means	7	7	7	8	7

Obtained from: Ramirez et al. (2004)

Table 4.6. Seasonal content of lignin (% dry matter) of native grasses collected in Teran county of the state of Nuevo Leon, México in 2001 and 2002

Native grasses	Seasons				Annual mean
	Winter	Spring	Summer	Fall	
Bouteloua curtipendula	8	8	7	8	8
Bouteloua trifida	9	8	7	9	8
Brachiaria fasciculata	9	7	9	8	8
Digitaria insularis	6	5	5	5	5
Chloris ciliata	8	7	8	7	8
Leptochloa filiformis	7	7	8	9	8
Panicum hallii	7	5	6	8	7
Panicum obtusum	8	7	8	9	8
Panicum unispicatum	6	8	7	6	7
Setaria grisebachii	9	8	8	8	8
Setaria macrostachya	8	9	8	9	9
Tridens eragrostoides	7	8	6	7	7
Tridens muticus	9	9	9	8	9
Seasonal means	8	9	6	6	7

Obtained from: Cobio-Nagao (2004).

CHAPTER 5

Nutrition of grass proteins

Introduction

Proteins are complex organic compounds of high molecular weight. Like carbohydrates and lipids, contain carbon, hydrogen and oxygen; in addition, all contain nitrogen and some have sulfur. Proteins are found in all living cells, where they are connected to all stages of life activities of the cells. Each species has its own proteins, and a simple organism contains different proteins in its own cells and tissues. Therefore, there are a large number of naturally occurring proteins. Proteins provide the amino acids needed for maintenance of vital functions, reproduction, growth and lactation. Nonruminant animals need preformed amino acids in their diets, but ruminants can utilize many other nitrogen sources because of their unique ability to synthesize amino acids and form protein from nonprotein nitrogen (NPN). This ability depends on the rumen microorganisms. In addition, ruminants possess a mechanism to save nitrogen: when the nitrogen content in the diet is low: urea a final product of protein metabolism in the body can be recycled in large quantities to the rumen.

Nitrogen

The nitrogen with symbol N and atomic number 7 is the fourth most ample compound in the biosphere. Nitrogen is a compound that is vital for all living organisms. Nitrogen is part of all proteins and it may be recognized in practically every living organisms. Nitrogen substances are existent in organic resources, feeds, organic composts, explosives and toxins. Life depends on N and is vital to, however, in excess; N to the environment may be detrimental.

Through the nitrogen cycle, the nitrogen in atmosphere is transformed to diverse groups of organic substances that represent the best important ordinary pathways to stand living creatures. During the N cycle, in the soil, bacteria process or trap the N from atmosphere transforming it into ammonia that plants require for tissue development (Figure 5.1). Nitrogen fixation, both artificial and natural, is indispensable for all systems of life due to N is used to synthesize the basic organs of plants, animals and other types of life, for example, amino acids for proteins and nucleotides for RNA and DNA. Thus, N fixation is indispensable for farming and the production of fertilizers.

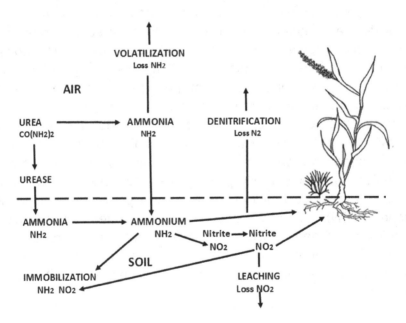

Figure 5.1. Nitrogen fixation by plants

Biological nitrogen fixation can include conversion to nitrogen dioxide. All biological nitrogen fixation is carried out by way of nitrogenase metalo-enzymes that contain iron, molybdenum, or vanadium. Microorganisms that can fix nitrogen are prokaryotes (both bacteria and archaea, distributed throughout their respective kingdoms) called diazotrophs. Some higher plants, and some animals (termites), have formed associations (symbiosis) with diazotrophs. Plants from the legume family that participate to N fixation comprise with groups such as clovers, alfalfa, soybeans, and peanuts, etc. These plants hold symbiotic bacteria that is named rhizobia inside the nodules in the roots, synthesizing N compounds that benefit the grow of plants and contend with other plants. When the plant terminates, the N fixed is liberated; creating it accessible to other plants and this aids to increase N in the soil. Most legumes have this relationship; however, only some genus do not have it. In some usual and organic farming systems, turfs are switched through several types of crops that mostly comprises one comprising chiefly or totally of clover or buckwheat (non legume) that are sometimes is mentioned as green fertilizer.

Metabolism of absorbed nitrogen

Few animals eat constantly, that means that the flow of nutrients in the body is sporadic, and inconsistent. The metabolic machinery must be prepared to support severe increases of nutrients, being able to temporarily store them in circulation during the stages of scarcity. The absorption and metabolism of N is no exception. For this process, the liver is the key organ of body that synthesizes proteins, provides circulation when amino acids are needed and processed for excretion when N exists excessively. Its proper functioning depends not only on its ability to absorb and retain amino acids, but also to provide adequate and careful release them to the whole system (Figure 5.2).

Nitrogen compounds

The N is a component of proteins and other compounds that are included in the organic matter of food (Figure 5.3). Proteins are composed of one or more chains of amino acids linked in peptide bonds. There are 20 amino acids found in proteins. The genetic code determines the structure of each protein that in turn provides a specific function in the body. Some amino acids are essential and non-essential. Nonessential amino acids can be synthesized in the body, but the essential amino acids must be present in the diet because the body cannot synthesize them.

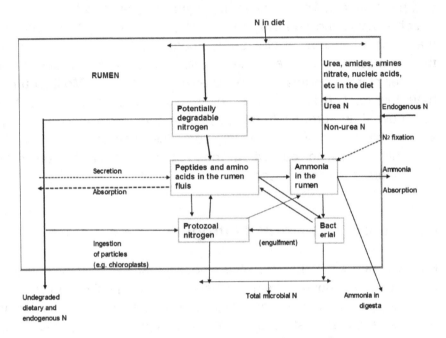

Figure 5.2. Absorption and metabolism of N in ruminants

The N in the food that is called non-protein nitrogen (NPN) is the N that is not found as part of the structure of a protein. Non-protein nitrogen (e.g. ammonia, urea, amines, nucleic acids, etc.) have no nutritional value for single-stomach animals. However, in ruminants, NPN can be used by rumen bacteria to synthesize amino acids and proteins that benefit the animal.

The Danish chemist J. G. Kjeldahl in 1883, developed a method to determine the amount of nitrogen in a compound. On average protein content of nitrogen is 16%. Thus, the percentage of protein in a food is typically calculated as the percentage of nitrogen multiplied by 6.25 (100/16 = 6.25). This measurement is called crude protein. It refers to raw word that not all nitrogen in the feed is in the form of protein. Usually the figure for crude protein gives an overestimate of the true percentage of protein in a food. Crude protein in forages is 5% (crop residues) to more than 20 % (good quality legumes). Animal by-products are usually very rich in protein (60% crude protein). Then the micro Kjeldahl technique is quite sensitive and provides a good estimate of the content of N as described.

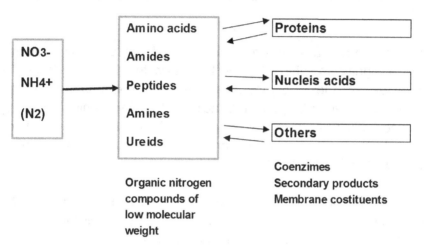

Figure 5.3. Organic nitrogen compounds

Nitrogen in forage

Plant protein is classified into two categories: 1) of leaves and 2) seeds. About half of the protein is soluble and is located in the chloroplasts, mitochondria and cytoplasm and more than 50% of the soluble leaf protein is composed of ribulose diphosphate carboxylase enzyme or Fraction I protein. The insoluble protein is associated with membranes, most of which is in chloroplasts. The amount of chloroplasts in the membrane increases with leaf age, only declines rapidly with senescence, with limited conditions such as low light intensity and sources of N, most chloroplast proteins are found in the membranes.

A large proportion (10-35%) of N in the forage is present as non-protein nitrogen (NPN) such as urea, free amino acids, amines, amides, nucleotides, peptides, chlorophylls and nonprotein amino acids, with small amounts of alkaloids and inorganic N. The NPN is higher in the immature forage and increases with N fertilization. The level of free amino acids decreases with age of the plant. In addition, there is a high difference for amino acids between leaves and stems. Nitrogen in leaves is higher than in stems.

Some of the non-protein amino acids found in grasses and legumes have a significant inhibitory effect or toxic for animals (Table 5.1). Examples of these compounds are: in *Leucaena leucocepha,* mimosine, in *Indigofera* species, acid α, β and γ-diaminobutyric and β-aminopropinitrilic, in *Lathyrus* species, S-methyl cysteine sulfoxide and in *Brassicas* containing Se amino acids in forage plants that are causing toxicity to livestock.

Nitrate is the only source of inorganic nitrogen in plants, it occurs mainly into the stems than in the leaves. While the nitrate in the ground is generally readily metabolized into other compounds, particularly NO3-N is accumulated in certain species due to N fertilization, drought or shaded conditions and herbicide application. High levels of NO3 on pasture, grassland products, and some brassicas can cause toxicity in ruminants because of its conversion to more toxic nitrates in the rumen.

Effect of the environment of N composition of plants

The analysis of a large amount of fodders from around the world shows an average crude protein content of 17% and 11.5 % legumes in grasses. Tropical grasses have less protein than temperate with small differences between populations of tropical legumes and temperate (values of 16.6 and 17.5%, respectively). The average crude protein of 369 perennial warm-season grasses collected in northeastern USA was 7.6%. There are significant differences in crude protein content between species of grasses and legumes, and among cultivars and within species (Table 5.1).

Table 5.1. Distribution of the protein in temperate forage leaves

Intracellular localization		% of total protein
Chloroplasts	- membrane	30-35
	- soluble	25-30
Mitochondrial	- membrane	3-4
	- soluble	2-3
Cytoplasm	- membrane	5-15
	- soluble	15-20
Nucleus		1-2
Cell wall		1-2

The effects of N fertilization on the content and composition of N in forages has been well documented. It has been reported that there was very little increase in the content of crude protein and water-soluble carbohydrates (WSC) with the application of 500 kg N ha^{-1} yr-1 from 29.5% to 30.4%, with and without application of N, respectively. In addition, the ratio of crude protein WSC changed from 1:1.1 are N fertilization to 1:0.5 with the application of 500 kg N. However, there were marked changes in the chemical composition of the silage and probably because the better efficiency of utilization of N and fiber by rumen microorganisms. The C4 tropical grasses typically have lower content of N species than C3 temperate grasses, but the former showed a better response to fertilization.

Many of the factors described so far that affect the content and form of N in forages are strongly influenced by climatic variables such as temperature, light and water. Higher temperatures increase the metabolic activity and the synthesis of structural compounds, causing decrease in NO3-N, WSC and protein in forages. Furthermore, the high temperature increases the formation of cell wall and decreases the digestibility of forage. The high light intensity increases the production WSC in plants, and reduction of shading promotes development of the cell wall; however, shaded plants generally have high protein content.

Most reports on the effect of drought or water supply showed that a moderate stress in the water supply, plant growth is reduced and therefore the crude protein content is increased, the fiber is reduced and digestibility is improved. However, the effect of climatic factors on the ground is difficult to predict because none of them operates independently of each other, since forage quality seems inconsistent from year to year. Traditional, cured hay produced in the sunshine resulted in low nutrient losses by breathing and draining of the cell contents caused by rain. Table 5.2 show the protein losses that occur by drying of fodder compared to the original feed.

Table 5.2. Crude protein content (%) of different types of hay from temperate grasses

Treatment	Environmental conditions		
	Dry season	Wet season	Average
Freezing at cut	11.9	9.4	10.6
Dry under shade	11.3	9.3	10.3
Dry in tripod	10.8	9.0	9.9
Rolled	9.4	8.6	9.0

Losses in crude protein were higher in the wet season compared to the dry season. A large number of methods, including treatment with propionic acid, ammonia and urea are usually used to maintain the hay reducing microbial activity and warming. Ammonification with anhydrous ammonia also increases the content of N in hay and improves its digestibility and intake. However, some researchers prefer the use of urea to preserve moist hay because of its low toxicity. Hay urea is rapidly converted by the plant urease into ammonia (NH3) reducing the temperature of the bale, increasing the content of N and improving the *in vitro* digestibility of dry matter and cell wall.

During the ensiled of forage, a large breakdown of leaf protein proteolysis occurs and continues until pH reaches 4, leading to accumulation soluble NNP. The main objectives of the process is to achieve anaerobic silage and suppression of clostridial growth due to the rapid formation of lactic acid. Silage additives function simulating lactic fermentation inhibiting microbial growth inhibiting anaerobic breakdown and providing nutrients in the form of nitrogen compounds or minerals.

The main constraints to animal feed straw or stubble are high levels of lignin in the cell wall and low levels of crude protein. Values of 4.0, 4.0, 3.0, 3.0 and 5.0% crude protein in wheat straw, corn stover, corn cobs and straw sorghum, respectively have been reported. Processing crop residues such as straw and stubble, with ammonia compounds are also effective, as described in the silage, to increase the N content and digestibility of the cell wall and voluntary intake.

Cereal grains are the main energy source in diets for ruminants, but they contain limited amounts of crude protein (10-13% dry basis). For grain processing, it has been used physical and chemical methods, as well as rolling and squeezing are used in combination with different forms of heat treatment. Ammonification as chemical treatment increased the concentration of N in the grain, and improved the *in vitro* protein digestibility compared to treatment with propionic acid.

Roles of proteins in the ruminant

Proteins makeup a vast and interesting group of organic substances. The enormous biological importance is due to the fact that cell protoplasm and all histological tissues, of any animal organism, contain proteins (Figure 5.4). Proteins are needed for milk production, muscle, skin, hair and to replace the inevitable loss of protein used in the maintenance of body weight. Proteins contain 22 different amino acids and ruminants may be unable to synthesize some of these amino acids in an amount sufficient to meet their optimal ratio requirements. These are defined as essential amino acids and must be absorbed after digestion in the small intestine.

There are many differences among the three tissues (milk, muscle and hair) in relation to the requirements of several essential amino acids; the muscle protein synthesis requires about twice the amount of arginine that milk protein. Hair protein synthesis requires very large amounts of sulfur amino acids as methionine, cysteine and cystine. In non-ruminant animals, the amount of essential amino acids available for absorption in the small intestine can be determined by analyzing the feed. This simple method cannot be applied to ruminants due to the presence of microbes in the rumen that modify the amount and proportions of amino acids available for absorption.

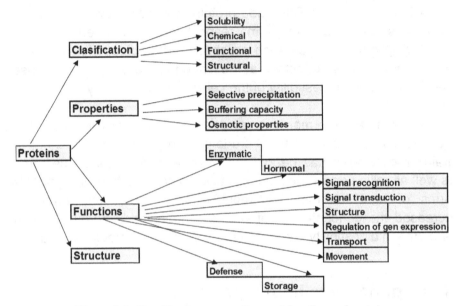

Figure 5.4. Classification, properties and functions of proteins

Three major changes occur for crude protein in the rumen: 1) degradation of crude protein into ammonia. If the ammonia produced is present in high concentrations, it is absorbed, leading to a net loss of crude protein, 2) microbial protein that is synthesized from NPN and sulfur present in the rumen. This can lead to a net gain of more crude protein entry into the duodenum, with high intake of crude protein, and 3) the microbes synthesize proteins that have a different profile of amino acids to proteins that are degraded in the diet.

The potential value of forage as amino acid generator may be determined by analysis of total nitrogen and sulfur. Forages contain nitrogen in many different forms and these can be converted into amino acids by rumen microbes. The term crude protein (CP) is used to describe all forms of nitrogen in the plant. Plants and animals contain amino acids, generally at an average of 160 g N/kg of dry matter.

Protein degradation in the rumen

The food proteins are degraded by microorganisms in the rumen, via amino acids, into ammonia and organic acids (multi-chain fatty acids). Ammonia also comes from sources of NPN in food and saliva urea recycled through the rumen wall. Low levels of ammonia causes a shortage of ammonia nitrogen for bacteria and reduces the digestibility of food. Too much ammonia in the rumen leads to weight loss, ammonia toxicity, and in extreme cases, death of the animal.

The amount of ammonia to synthesize microbial protein depends mainly on the availability of energy generated by the fermentation of carbohydrates. On average, 100 g organic matter fermented in the rumen are required to synthesize 20 g of bacterial protein. The bacterial protein synthesis can vary from 400 g to about 1500 g^{-1} day^{-1} depending of diet digestibility. The percentage of protein in bacteria varies from 38 to 55%.

A primary factor controlling the rate of release of ammonia by degradative deamination in the rumen is the solubility of the protein, factor that is controlled by physical and chemical characteristics. The solubility of proteins found in natural foods varies considerably and can be determined in the rumen following the increasing of the ammonia concentration in the rumen fluid.

Protein in feces

Almost 80 % of the protein reaching the small intestine is digested and the rest goes to the feces. Another important sources of nitrogen in the feces are the digestive enzymes secreted by the intestine, and the rapidly replacement of the intestinal cells (metabolic protein feces). On average, for every increase of 1 kg of dry matter intake by the ruminant, there is an increase of 33 g of body protein lost in the intestine and excreted in the feces. Ruminant feces are good fertilizer because they are rich in organic matter and is particularly rich in nitrogen (12.2-12.6% N or the equivalent of 14-16% crude protein) as compared to the feces of non-ruminant animals.

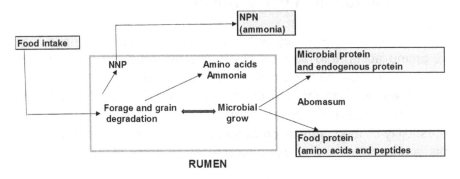

Figure 5.5. Ruminal digestion of proteins

Classification of proteins

Proteins are classified according to the type, shape, structure, solubility and/or chemical composition. The most common types of protein in grass include:

1) Albumins that are soluble in water and insoluble in alcohol,
2) Globulins that are insoluble in water and alcohol, but soluble in salt solutions,
3) Prolamins that are soluble in alcohol, but insoluble in water and saline solutions, and
4) Glutelins that are soluble only in alkaline solutions.

While many seeds may contain varying fractions of different types of proteins, cereal grains tend to contain large amounts of prolamins and glutelins, as the insoluble and hydrophobic characteristics of corn and wheat proteins, which have low rates of ruminal hydrolysis. Dicots, legumes particularly, tend to contain globulins and albumins potential soluble, so they are more sensitive to heat denaturation, giving them water insolubility.

Nutritional quality of microbial protein

The amount of protein reaching the small intestine for absorption is the amount of protein synthesized by rumen microbes and dietary protein escaping (escape protein) to the rumen, being referred to by the NRC (2000) as bacterial crude protein (BCP) and non-degradable intake protein (NDIP), respectively. Cattle and other ruminants can degrade fibrous feedstuffs owing to the consortium of bacteria, protozoa, fungi, and methanogens inhabiting their rumen and hindgut. This consortium ferments fiber and other feed components to short-chain fatty acids (SCFA) and, in the process, generates ATP that fuels microbial growth (synthesis of cellular protein in particular). This microbial protein supplies 60 to 85% of amino acids (AA) reaching the animal's small intestine. Maximizing efficiency of its production would consequently improve cattle productivity, and the amount depends on factors such as diet or the animal itself.

Some definitions that are be relevant to understand microbial protein concept are:

1) Rumen degradable protein (RDP) is defined as that portion of dietary protein that can be degraded in the rumen, the largest of the multi-compartmental stomach, by microorganisms (both bacteria and protozoa) that use the protein to manufacture high quality microbial cell proteins, also known as microbial crude protein (MCP).

2) Rumen undegradable protein (RUP) is defined as that portion of dietary protein that escapes degradation by ruminal microorganisms and is passed into the small intestine for digestion and absorption. The proportion of total feed protein that is undegradable is not constant from one feedstuff to another. Although frequently referred to as bypass proteins, they technically do not bypass the rumen, but are simply not utilized by the microorganisms as a substrate to make MCP.

3) Metabolizable protein (MP) is defined as the true protein absorbed in the small intestine and is composed of RUP and MCP.

The quality of the protein and the essential amino acid requirements were obtained originally, from the requirements of no-ruminants species, but then were withheld in ruminants; in addition, some studies indicated that differences in dietary protein quality for ruminants might not be detected. However, with the development of new methods, such as the measurement of the flow of intake was possible to subdivide the total digestion into digestive absorption activities on the reticulo-rumen, abomasum, small intestine and large intestine. In addition, with techniques such as feeding by parenteral infusion with mixtures of amino acids (AA) of known composition, you can determine more accurately patterns of essential amino acid requirement for ruminants. Moreover, the rumen *in vitro* gas production technique used for feed energy evaluation in ruminant feedstuffs was adopted to assess microbial biomass production potential of feedstuffs based on the concept of partitioning of fermented organic matter between microbial biomass and fermentation waste products.

Actually, there is limited knowledge regarding the quantitative requirements of AA in ruminants, since the microbial involvement inhibits a correct study. During ruminal digestion, the microbes are multiply and synthesize in considerable amounts, producing microbial protein using NNP and dietary protein as sources of N. The concept of NNP as a source of N to the rumen microbes was based on the results of the first investigations, which suggested that dietary protein is fermented in the rumen into simple nitrogen compounds and are reincorporated to bacterial cell protein as NH_3 N. Eventually, it was shown that cattle fed NNP can grow and reproduce without feeding AA or dietary protein.

When compared, the composition of the essential AA (EAA) and semi-essential AA (SEAA) contained in several commons foods for ruminants, it was concluded that the microbial protein is a suitable source of EAA (Table 5.3). Although the results of the above study suggest that a diet low protein quality would be improved if the protein was degraded in the rumen and converted to microbial protein; however, it was found that in high-production ruminants, the microbial protein alone may be insufficient to meet the demands of these animals for the production of animal protein. Patterns EAA and microbial protein are not markedly affected by diet, but the quality of the protein is not ideal, as its biological value is 66-87. Therefore, the growth of rapid production ruminants depends on the synthesis of microbial protein and dietary protein escaping from ruminal digestion to provide EAA.

Table 5.3. Comparisons between the amino acid profile of milk, rumen bacteria, and estimated rumen undegradable protein fractions of common protein sources

Amino acid	g/100 grams of amino acids					
	Milk	Bacteria	Blood milk	Canola Meal	Corn gluten	Fish meal
Arginine	3.4	5.1	4.1	5.0	3.3	5.7
Histidine	2.6	2.0	6.3	2.0	1.9	2.0
Isoleucine	5.8	5.7	5.3	3.2	3.8	2.7
Leucine	8.3	8.1	12.8	7.8	18.1	7.1
Lysine	7.5	7.9	8.8	5.1	2.0	7.0
Methionine	2.5	2.6	1.1	1.9	2.6	3.0
Phenylalanine	4.6	5.1	6.6	4.1	6.6	3.8
Threonine	4.4	5.8	4.3	4.7	3.5	4.3
Valine	6.3	6.2	7.5	4.0	4.3	3.3

Obtained from: National Dairy Council, 2000.

Urea

Numerous NPN compounds have been used as ingredients of livestock feeds. All compounds proved useful sources of nitrogen except biuret, but none exceeded urea in availability. The urea from feed sources is rapidly dissolved and hydrolyzed to ammonia by bacterial urease in the rumen. The ammonia can then be utilized by the bacteria for synthesis of amino acids required for their growth. Amino groups are also split from amino acids and from intact proteins and used by bacteria in the same manner. Protein synthesis within the rumen by microorganisms is very closely associated with the activity of these same organisms in breaking down cellulose and other carbohydrate materials and in the formation of organic acids as byproducts of this fermentation process. The solubility of natural proteins vary greatly and thus the rate at which they are hydrolyzed and utilized by bacteria differs appreciably. When ammonia is produced too rapidly in the rumen or if the concentration becomes too high, appreciable amounts are absorbed directly into the bloodstream, reconverted to urea in the liver, excreted through the kidneys in the urine and thus lost to the animal. There is, however, always a small amount of urea in the bloodstream and other body fluids. This urea finds its way into the saliva and re-enters the rumen. Urea has been shown to pass into the rumen directly through the rumen wall from the circulating blood.

There are diverse physiological functions of nitrogen products in different animal groups, including excretion, acid–base regulation, osmoregulation and buoyancy. There are physiological, biochemical and evolutionary characteristics that make animals to be ureotelic (excrete urea), ammonotelic (excrete ammonia) or uricotelic (excrete uric acid). Animals excrete a variety of nitrogen waste products, but ammonia, urea and uric acid predominate (Figure 5.9). A major factor in determining the mode of nitrogen excretion is the availability of water in the environment. Generally, aquatic animals excrete mostly ammonia, whereas terrestrial animals excrete either urea or uric acid. Ammonia, urea and uric acid are transported across cell membranes by different mechanisms corresponding to their different chemical properties in solution. Ammonia metabolism and excretion are linked to acid–base regulation in the kidney, but the role of urea and uric acid is less clear. Both invertebrates and vertebrates use nitrogen-containing organic compounds as intracellular osmolytes. In some marine invertebrates, NH_4^+ is sequestered in specific compartments to increase resilience.

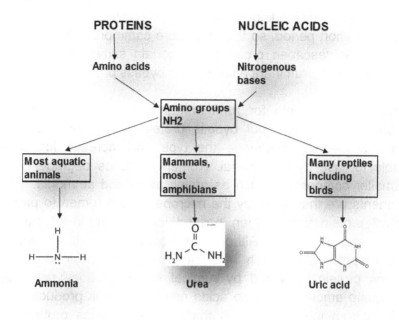

Figure 5.6. Ammonia. Urea and nucleic acid

Escaped and bypass proteins

Not all protein sources are processed the same way in ruminant animals. Some proteins are degraded in the rumen and are made in to bacteria and the bacteria become the source of protein in the small intestine for the animal. However, some proteins pass through the rumen undigested and are primarily digested and absorbed by the lower digestive tract. When this occurs, the ruminant animal has more total protein available for metabolism as needed to meet nutritional demands. These proteins can be used efficiently and cost effectively to feed ruminant animals if production demand exceeds the rumen's ability to meet the animals' nutritional needs.

Bypass proteins, also known as protected proteins or escape proteins, are the proteins that escape degradation in the rumen and pass on to the abomasum and small intestine for digestion and absorption. The term "bypass" is misleading because these proteins do reside in the rumen for a short period, so they are more commonly and accurately referred to as "escaped proteins". In certain situations, there is a benefit in providing ruminant animals with bypass proteins.

The majority of protein for ruminant animals is protein made by microbes in the rumen, but bypass protein in supplemental feeds provides a method to change the type and increase for protein available for use by the animal. Rumen microbes degrade protein into smaller components such as amino acids and ammonia. These components are then used by the microbes in the rumen to produce microbial protein that is then passed into the small intestine for digestion and absorption. Although microbial protein is a beneficial source of protein, research has shown that in certain production situations protein produced by ruminal synthesis does not supply an adequate amount of amino acids needed for milk production and growth by ruminant animals. In these cases, bypass proteins can be supplemented to provide the protein needed to bridge the gap between the protein needed by ruminant animals and the amount of microbial protein produced by rumen synthesis. The key focus of protein supplementation is increased nutrition at a lower cost. This is accomplished by maximizing the intake of less expensive non-protein diet-dependent nitrogen used as a source of ammonia for rumen synthesis, optimizing the intake of natural proteins highly degradable in the rumen, and the bypassing of proteins from the rumen to the lower digestive tract if more total protein is needed.

During the periods of growth and lactation, protein requirements significantly increase. In these cases, some ruminant animals may not be capable of producing enough total protein or some essential amino acids needed to meet their nutritional needs. Either an increase in the amount of protein fed is required, or a more efficient protein is needed to bypass degradation in the rumen, preserving digestion and absorption in the small intestine. Proteins sources have different bypass rates. Urea is a nitrogen source with no bypass value. Soybean meal has a bypass rate of 20%-40%. Proteins with medium bypass rates (40%-60%) include cottonseed meal, dehydrated alfalfa, corn grain, and brewers dried grains. Proteins with high bypass rates (60% +) include corn gluten meal and distillers dried grains.

Many factors can influence the bypass protein content of a forage. Listed below are common factors most often associated with creating variation in forage bypass protein content.

1. *Maturity.* Several studies have demonstrated that immature forage legumes and grasses contain more degradable and less undegradable protein than mature forages. Immature forages contain more NPN primarily composed of ammonia, nitrate, amines, amides, and free amino acids that are rapidly degradable in the rumen. With advancing maturity, true plant protein synthesis advances and the cell wall matrix becomes more complex, rendering forage protein less accessible to rumen bacteria and less degradable. These factors ultimately reduce degradation potential of forage proteins. Thus, maturity has a profound and large influence on bypass protein content of forages.

2. *Species.* Species is also known to affect bypass protein content of forages. In general, legume protein is more degradable than grass protein. This is due in part to grasses containing more neutral detergent fiber that reduces rates of nutrient digestion. Within grasses, bromegrass and quackgrass appear to have greater levels of bypass protein, while perennial ryegrass protein appears to be quite degradable. In addition, varieties within a forage species have been demonstrated to vary in bypass protein content.

3. *Fertilization.* Grasses assimilate soil nitrate (NO_3) and ammonium (NH_4) into NPN and true protein fractions. Increasing soil N supply increases forage N (crude protein). The increase in forage N (crude protein) is, however, disproportionate with the NPN pool increased largely than the true protein pool. Because NPN is readily degradable in the rumen, nitrogen fertilization generally reduces the amount of bypass protein. Fertilizing grasses with 0 or 60 kg/ha of N resulted in increasing crude protein 1.5% units and increasing protein degradability.

4. *Ensiling (Proteolysis).* When forages are ensiled, bacteria ferments the forage and breaks forage protein down into smaller fractions which are more degradable by rumen bacteria. This process is called proteolysis. It has been

estimated that only 9% of forage macro protein molecules remain after fermentation. The effect of proteolysis can have a dramatic effect on the bypass protein content of forages. The concept of proteolysis has been demonstrated from a recent study when alfalfa silage was made at three different maturities and wilted for 0, 10, 24, 32, 48, and 54 hours before ensiling. Ruminal degradability of ensiled forages was compared to a non-ensiled forage. In all cases, the percent crude protein remaining (bypass) was less for the ensiled forages as compared to the non-ensiled forages. The effects of reduced bypass protein content were also more pronounced as wilting time decreased (increased moisture content). Red clover silage is a notable exception to the conceptual effects of proteolysis on ruminal protein degradation. Red clover contains polyphenol oxidases that have been demonstrated to inhibit or reduce proteolysis during fermentation. Therefore, red clover silages will generally have a higher bypass protein content than alfalfa or grasses silages within a similar maturity and/or nutritional level.

5. *Heat Damage.*- When forages are ensiled too dry and/or elimination of oxygen from the silage mass is not satisfactory, significant levels of heat can be produced during the fermentation process. Significant levels of heating can also occur when legume or grass hays are made too wet. In these situations when excessive heating occurs, forage protein may become bound (maillard reaction) to forage carbohydrate fractions, rendering the protein fraction less degradable. Few studies have quantified these effects under field conditions. Controlled research, however, has clearly demonstrated the effects of heated forages on ruminal protein degradability. Studies demonstrated that when heated silages at 100 °C for just 2 min and observed a 6.0 percentage unit increase (18.9 vs. 24.5% of CP) in the amount of bypass protein in alfalfa silage.

Protein content in cultivated grasses

The content of crude protein (CP) in samples of cultivated grasses and introduced to northeast of Mexico varied from

7.7 to 11.9% dry matter (DM) depending on season, year, collection site and soil fertility level, with an average of 9.4% (Table 5.4.). These percentages are similar to CP average value (10.0%) of 560 tropical grasses, but are lower than the average value (16.6%) of 340 tropical legumes, 470 temperate grasses (13.3%) and 270 native trees and shrubs (17.0%). When the CP content of a grass is less than 6-7%, the ruminant intake will decrease because the low level of CP. This effect is due to microbial growth in the rumen is limited by CP deficiency. Therefore, cultivated pastures, non-irrigated and fertilized, growing in northeastern Mexico (Table 5.4.) Only during the summer and fall, contain levels of CP in sufficient amounts to meet the demands of CP for maintenance and production of ruminants growing. In these regions, the highest rainfall occurs during the summer and fall, thus this may be the reason why evaluated grasses of Table 5.4 are higher in CP content during those seasons.

Apparently, the hybrid buffel nueces and five new genotypes of buffelgrass (Cenchrus ciliaris), considered of high dry matter production, that were planted under rainfed conditions in Teran, Nuevo Leon, Mexico, during the summer of 1999 and autumn of 1999 and 2000 (Table 5.5) had CP levels of 7.0, 6.2 and 8.3%, respectively. Only in November 2000, the genotypes had marginally higher levels of CP to meet the demands of maintenance of growing grazing ruminants.

It has been proved that the crude protein content of grasses also depends on the availability of N in the soil; the fertilization with nitrogen generally increases the percentage of PC in the grasses. Apparently, the application of 120 kg of urea ha-1 as fertilizer, during June 2000, to the genotypes of buffelgrass shown Table 5.5, had a beneficial effect since it caused an increase in the average percentage of CP from 7.0 (August 1999, November 1999 and November 2000) to 8.7%.

On the other hand, CP levels ranging from 7-8% dry matter were reported (Table 5.6) in 84 new genotypes of buffelgrass planted under rainfed conditions and unfertilized, in Teran, NL, Mexico in November 2000. The CP was higher in genotypes identified as 409220 (9.0%), while 364445 was the lowest (6.0%). The average was 7.7%. This value is similar to other reports that evaluated genotypes of buffelgrass collected during autumn in the northeastern region of Mexico.

Table 5.4. Seasonal content of crude protein (% dry matter) of introduced grasses collected in different places of the state of Nuevo León, México

Grasses	Place and date of collection	Seasons and date of collection				
		Winter	Spring	Summer	Fall	Mean
Cenchrus ciliaris, Común	Marín, N.L., México (1994)	5.7	6.1	17.0	7.2	9.0
Cenchrus ciliaris, Común	Marín, N.L., México (1998)	6.8	6.3	12.7	8.8	8.7
Cenchrus ciliaris, Común	Terán, N.L., México (2001-02)	9.7	9.8	14.6	12.7	11.7
Cenchrus ciliaris, Común	Linares, N.L., México (1998-99)	5.7	10.9	14.0	10.9	10.4
Cenchrus ciliaris, Nueces	Marín, N.L., México (1998-99)	5.1	4.4	16.6	4.5	7.7
Cenchrus ciliaris, Llano	Marín, N.L., México (1998-99)	4.4	5.5	12.3	5.5	6.9
Cynodon dactylon (Cross I)	Linares, N.L., México (1998-99)	8.5	10.9	16.5	16.2	13.0
Cynodon dactylon (Cross I)	Marín, N.L., México (1994)	18.3	10.4	11.0	13.2	13.2
Cynodon dactylon (Cross II	Marín, N.L., México (1994)	12.0	10.3	9.6	10.1	10.5
Dichanthium annulatum	Linares, N.L., México (1998-99)	6.9	7.8	9.7	9.3	8.4
Dichanthium annulatum	Marín, N.L., México (1994)	5.5	4.9	4.2	4.6	4.8
Panicum coloratum	Linares, N.L., México (1998-99)	6.6	10.7	11.7	9.6	9.7
Rhynchelytrum repens	Terán, N.L., México (2001-02)	9.3	7.4	11.2	11.1	9.8
Promedio estacional		**6.9**	**8.1**	**13.5**	**9.6**	**9.5**

Obtained from: Ramírez et al. (2002); Ramírez et al. (2003); Ramírez (2003); Ramírez et al. (2004); Ramírez et al. (2005);

Table 5.5. Crude protein content (% dry matter) of the hybrid Nueces of buffelgrass and five genotypes of buffelgrass (Cenchrus ciliaris) collected in different places of Teran, Nuevo Leon, México

Genotypes	Collection dates			
	August 1999	November 1999	November 2000	June 2000 (Fertilized)
Cenchrus ciliaris Nueces	7.8	6.9	8.4	8.8
Cenchrus ciliaris 307622	6.3	5.1	7.8	8.0
Cenchrus ciliaris 409252	6.5	6.7	8.1	8.5
Cenchrus ciliaris 409375	7.7	5.6	8.8	8.1
Cenchrus ciliaris 409460	6.1	6.5	8.2	9.6
Cenchrus ciliaris 443	7.4	6.2	7.5	9.1
Mean	7.0	6.2	8.0	8.7

Obtained from: García-Dessommes et al. (2003ab)

Protein in leaves and stems

The CP in grasses generally decreases as they mature. This decrease is due to an increase in the stem proportion. This is shown in the data in Table 5.7; without exception, in all grasses growing in northeastern Mexico, the leaves had higher CP content than stems. Therefore, in grasses, the leaves are usually the highest quality part of the plant. In grasses, the leaves comprise blade and sheath. In dicotyledonous plants like legumes, leaves consist of lamina and petiole. These components differ in leaf quality and should not be considered as a single entity. It has been reported that cattle and sheep consumed 35 to 21% fraction of the leaf fraction pangola stem grass (*Digitaria decumbens*) and Rhodes grass. Both fractions, stem and leaf were digested in the same extent; therefore, increased consumption was attributed to the short time that the leaf fraction was retained in the rumen compared to stem fraction. Moreover, in a number of studies around the world, it was concluded that intake of leaves was higher than intake of stems.

Table 5.6. Crude protein content (CP, % dry matter) of 84 new genotypes of *Cenchrus ciliaris* collected in Teran, N.L., México in November 2000

Genotypes	CP, %	Genotypes	CP, %	Genotypes	CP, %	Genotypes	CP, %
202513	7	409185	8	409258	7	409449	8
253261	8	409197	8	409263	8	409459	8
307622	8	409200	8	409264	7	409460	9
364428	8	409219	8	409266	8	409465	8
364439	7	409220	9	409270	7	409466	7
364445	6	409222	8	409278	7	409472	8
365654	7	409223	8	409280	8	409480	8
365702	7	409225	8	409300	8	409529	8
365704	8	409227	8	409342	8	409691	8
365713	8	409228	8	409359	8	409711	7
365728	8	409229	8	409363	8	414447	7
365731	8	409230	8	409369	7	414451	8
409142	7	409232	8	409373	8	414454	8
409151	7	409234	8	409375	7	414460	7
409154	7	409235	7	409377	7	414467	8
409155	8	409238	8	409381	8	414499	8
409157	7	409240	8	409391	8	414511	8

409162	8	409242	8	409400	8	414512	8
409164	7	409242	8	409410	8	414520	8
409165	7	409252	8	409424	8	414532	8
409168	8	409254	7	409448	7	443	8
Mean of all genotypes							8

Obtained from: Rodríguez-Morales (2003)

Protein in native grasses

Among species of grasses, CP content is highly correlated with many of the nutritional attributes of plants as their digestibility, vitamins, Ca and P content; however, all decline in CP to deficient levels at the same time; therefore, the CP serves as a general measure of acceptable nutritional quality of grasses. If it is considered that 7.0% is an appropriate level of CP for maintenance in ruminants, with exception of native grasses such as *Aristida longiseta* and *Hilaria belangeri*, all grasses that are shown in Tables 5.8 and 5.9 can be considered of good nutritional quality, in all seasons, for growing ruminants.

Table 5.7. Crude protein content (% dry matter) in leaves and stems in cultivated grasses collected in different places and dates of the state of Nuevo Leon, Mexico

| Grasses | Place and date of collection | Parts | Seasons | | | | Annual mean |
			Winter	Spring	Summer	Fall	
Cenchrus ciliaris Common	Linares, N.L., México (1998-99)	Leaves	8.3	13.8	13.1	17.3	13.1
		Stems	3.6	8.9	8.5	10.3	7.8
Cenchrus ciliaris Common	Marin, N.L., México (1998-99)	Leaves	7.2	6.6	18.7	9.9	10.6
		Stems	4.5	6.0	12.6	4.3	6.9
Cenchrus ciliaris Llano	Marin, N.L., México (1998-99)	Leaves	4.9	5.4	15.7	8.5	8.6
		Stems	3.1	2.7	9.6	4.1	4.9
Cenchrus ciliaris Nueces	Marin, N.L., México (1998-99)	Leaves	5.1	2.9	17.5	5.5	7.8
		Stems	2.7	2.2	12.8	3.4	5.3
Cynodon dactylon	Linares, N.L., México (1998-99)	Leaves	10.4	21.0	19.3	13.8	16.1
		Stems	5.6	12.6	13.3	8.3	10.0
Dichanthium annulatum	Linares, N.L., México (1998-99)	Leaves	5.4	12.3	8.2	12.9	9.7
		Stems	2.5	7.3	7.4	5.2	5.6
Panicum coloratum	Linares, N.L., México (1998-99)	Leaves	4.6	13.7	9.4	11.7	9.9
		Stems	3.2	7.1	6.1	5.5	5.5

Obtained from: Ramírez *et al.* (2001ab); Foroughbackhch *et al.* (2001); Ramírez *et al.* (2003); Ramírez (2003); Ramírez *et al.* (2005).

All native grasses exhibited its highest percentage of CP in the summer season compared to other seasons. Seasonal fluctuations of CP content may be due to the summer rainfall. Other studies that evaluated the protein content in native grasses, also showed seasonal fluctuations. It is reported that *H. belangeri* had a value of 13% in summer and decreased to 2.0 % in winter. Also in B. gracilis, collected in Wyoming, USA, the CP had values of 11 % in spring, but 6.0% in winter. Studies conducted in Sonora, Mexico reported that B. gracilis, and Aristida spp and Setaria macrostachya during the winter and summer of 1989 had CP values of 5 and 10%, 5 and 9%, 7 and 10%, respectively. However, in another study, it was reported that P. hallii harvested in Texas, USA, reported similar CP values of 7% in summer and autumn.

Table 5.8. Crude protein content (% dry matter) of native grasses collected in Marin, N.L., Mexico in 1994

Native grasses	Seasons of the year				Annual mean
	Winter	Spring	Summer	Fall	
Aristida longiseta	5.9	5.6	6.2	5.9	5.9
Bouteloua gracilis	8.6	6.7	10.5	9.1	8.7
Cenchrus incertus	5.9	7.2	9.5	9.6	8.1
Hilaria belangeri	4.4	6.6	8.1	9.4	7.1
Panicum hallii	9.3	13.0	14.2	12.2	12.2
Setaria macrostachya	7.6	10.5	10.9	9.4	9.6
Seasonal means	7.0	8.3	9.9	9.3	8.6

Obtained from: Ramirez *et al.* (2004)

Proteins are composed of amino acids. In general, the amino acids glutamic acid, aspartic acid and arginine are the major of the protein components in grasses. In addition, the concentration of CP in grasses is mainly influenced by the supply of available soil N and the state of maturity of the plants. As previously mentioned, in grasses CP concentration severely declines as maturity increased, possibly due to the relative increase of the cell wall and cytoplasmic decrease. It is likely that this effect became clear during the winter and spring pastures listed in Tables 5.8. and 5.9, respectively because at these seasons was reported the lowest values of CP, especially in Bouteloua curtipendula and Bouteloua trifida, but not in Brachiaria fasciculata that had the highest level of CP, along with Panicum obtusum.

Protein requirements in ruminants

Ration formulation to meet the requirements of protein and amino acids in ruminant systems were developed based on digestible crude protein calculations. However, at the present it is necessary to express the protein requirements in terms of metabolizable protein (MP). These methods take into account the impact of rumen fermentation of the protein components in the diet, and take account the N requirements for the synthesis of microbial protein in the rumen. The key components of the MP of foods are summarized in Table 5.10.

Table 5.9. Seasonal crude protein content (% dry matter) of native grasses collected in the county of Teran of the state of Nuevo Leon, México in 2001 y 2002

Native grasses	Seasons and year of collection				Annual mean
	Winter	Spring	Summer	Fall	
Bouteloua curtipendula	9.1	8.2	11.9	13.7	10.7
Bouteloua trifida	7.9	7.9	14.8	13.0	10.9
Brachiaria fasciculata	10.8	10.0	16.3	18.4	13.9
Digitaria insularis	11.3	7.5	13.3	13.2	11.1
Chloris ciliata	11.7	10.0	18.2	14.6	13.6
Leptochloa filiformis	11.0	9.8	12.0	15.1	12.0
Panicum hallii	10.6	8.0	16.1	17.6	13.1
Panicum obtusum	11.9	12.6	16.5	14.5	13.9
Panicum unispicatum	8.7	8.9	13.4	13.2	11.1
Setaria grisebachii	8.8	13.2	17.1	15.1	13.5
Setaria macrostachya	11.0	12.3	15.5	11.4	12.5
Tridens eragrostoides	12.4	12.5	16.6	11.1	13.2
Tridens muticus	8.0	8.3	15.9	12.2	11.1
Seasonal means	10.2	9.9	15.2	14.1	12.4

Obtained from: Cobio-Nagao (2004)

Table 5.10. Components and factors determining the metabolizable protein components

Components	Factors that affect the MP content of foods
Degraded protein in rumen	Solubility (fraction rapidly degraded/), ruminal secretion index and feed level
Synthesis of microbial protein	Energy supply inside the rumen and N supply inside the rumen
Undegraded digestible protein	Reach of degradability in rumen, proportion of insoluble N (e.g. bound lignin) and amino acid composition of material that escaped the rumen
Use of amino acids	Absorption efficiency and utilization efficiency for maintenance growth, milking and pregnancy

Metabolizable protein (MP) is the total amount of amino acid absorbed in the small intestine. The main sources of amino acid to the intestine are rumen-undegradable protein (RUP), microbial crude protein (MCP), and to a lesser extent, endogenous crude protein (ECP). The absorbed amino acids may be used in several metabolic processes, such as milk production, immune system, reproduction, etc. Measurement of RDP and RUP can be performed *in situ* or *in vitro*, and measurement of microbial crude protein requires a more elaborate technique, sampling content from the rumen, abomasum and small intestine, and using external markers. Balancing rations for RDP, RUP and MP leads to an improvement in nitrogen efficiency, which has a direct effect on nitrogen excretion, feed costs and performance. In order to optimize production with the least-cost diet, it is fundamental to know the RDP and RUP for each feedstuff as well as the animal's requirement for each amino acid at different stages of production.

CHAPTER 6

Digestibility of grasses

Introduction

Digestion is the degradation of macromolecules of food into simpler compounds that can be absorbed in the gastrointestinal tract. Digestion in mammals takes place through two strategies. In a first stage, digestion is mediated by acid hydrolysis in the stomach and digestive enzymes in the small intestine. In the second stage, digestion occurs by a fermentative metabolism of dietary components held by microbes that occupy certain compartments in the digestive tract. Actually, all mammals practice both types of digestion, but the degree of emphasis on either strategy varies considerably. Herbivores species are characterized by the presence of extensive fermentation site in the alimentary tract. In some animals (e.g. horses), the main fermentation takes place in the large intestine (cecum and colon). In others, the fermentation occurs mainly on a modified four-compartment stomach. Ruminants, of course, are animals that have achieved their maximum fermentative capacity on the front stomach. Although ruminants are characterized because the fermentation occurs mainly in the rumen, the digestion that is carried out in abomasum and small intestine are also processes of vital importance to the ruminant.

Structure and functionality of the digestive system of ruminants

The rumen is one of four stomach compartments found in ruminants. Ruminants are animals such as cattle, sheep, goats and deer. (In comparison, animals such as pigs, dogs and horses have only a single stomach compartment and are called nonruminants, or monogastric animals.) The rumen allows grazing animals to digest cellulose, a very common carbohydrate in plants. Three of the four ruminant stomach compartments make up the forestomach. These three compartments (rumen, reticulum, and omasum) are an extension of the lower esophagus. The rumen, the first of the forestomach chambers, stores and processes plant material. It can be a very large structure indeed: in large ruminants, the rumen may store up to 95 l of undigested food. The rumen holds plant material until it has been broken down, releasing volatile fatty acids, and fermentation of protein and carbohydrates has begun.

Ruminant digestion begins when the animal swallows a mouthful of plants. The food is partially chewed and mixed into a bolus with saliva, before being swallowed and passing down the esophagus into the rumen. When ruminants are grazing they tend to swallow their food quickly, with only minimal mastication. When the animal is resting after grazing, it regurgitates this partially chewed food, rechews it, and then swallows it again (rumination.) Depending on the amount of fiber in their food, cattle may spend between 3 to 6 hours per day chewing.

Mouth. It is the lobby of the digestive system. It is understood cavity between the jawbone and palatine, elongated along the axis of the head, and two openings, anterior and posterior.

Esophagus. Is a muscular tube that connects the mouth with the forestomach. Food passes down the esophagus by contraction of the muscles in the walls that push the food along in a series of waves called peristalsis. The ruminant esophagus is also capable of reversed peristalsis or antiperistalsis. This allows food to be easily regurgitated from the rumen and chewed.

Reticulo-rumen. Is composed of the rumen and the reticulum. The reticulo-rumen is partially separated from the rumen by the reticular fold, which allows mixing between the two compartments. The contents of the reticulo-rumen are mixed by contractions of the reticulo-rumen wall. The mixing recirculates undigested material preventing the rumen becoming clogged and distributing symbiotic bacteria throughout the ingested material. The reticulo-rumen becomes colonized by symbiotic bacteria in the first week after birth. The bacteria help to break down the food and release nutrients by a fermentation process.

Omasum. When food has been broken down enough, it passes from the reticulo-rumen through the reticulo-omasal orifice to the omasum. The omasum wall is highly folded, giving a large surface area that allows for the efficient absorption of water and salts released from the partially digested food. The omasum also acts as a type of pump, moving the food from the reticulorumen to the true stomach, the abomasum, where acid digestion takes place.

Abomasum. Unlike a ruminant's three forestomachs, the abomasum is a secretory stomach. This means that cells in the abomasum wall produce enzymes and hydrochloric acid that hydrolyze proteins in the food and in the microbes mixed in with the food. Hydrolysis breaks the proteins into smaller sub-units (e.g. dipeptides and amino acids), ready for further digestion and absorption in the small intestine. Because ruminants eat such large amounts of plant material, there is an almost continuous flow of food through the abomasum. In comparison, activity in the stomach of nonruminant animals generally has a circadian rhythm associated with food intake.

Small intestine. Is an elongated tube running from the abomasum to the large intestine. In ruminants, the small intestine is about 20 times longer than the length of the animal. Thus, a cow two meters in length would have a small intestine of 40 meters long. A large proportion of the digestion and absorption of nutrients and water occurs in the small intestine. Enzymes in the small intestine break nutrient molecules down into their building blocks. Carbohydrates are broken down to simple sugars, fats into fatty acids and monoglycerides, nucleic acids into nucleotides and proteins into amino acids. Some of these enzymes are on the surfaces of intestinal cells, while others are secreted into the small intestine, primarily from the liver and pancreas. The small intestine has three regions: the duodenum, the jejunum and the ileum. Partially digested food passes from the duodenum along the small intestine by way of peristaltic muscle contractions that start at the part where the abomasum is joined to the duodenum.

Duodenum. The liver and pancreas both secrete materials through ducts into the duodenum. The common bile duct carries bile salts, a greenish fluid that is manufactured in the liver, stored in the gall bladder (the ruminant gall bladder does very little to concentrate the bile), and released into the duodenum to digest fats. The main pancreatic duct carries digestive secretions, which are rich in enzymes and bicarbonate. The bicarbonate neutralizes acid from the stomach, which would otherwise inactivate many of the duodenum's digestive enzymes.

Jejunum. The lining of the jejunum is specialized for the absorption of carbohydrates and proteins. Its inner surface is covered in finger-like projections called villi, which increase the surface area available to absorb nutrients from the gut contents. The villi in the jejunum are much longer than in the duodenum or ileum. The epithelial cells that line these villi possess even larger numbers of microvilli, known collectively as the brush border. The combination of villi and microvilli increases the surface area of the small intestine, increasing the chance of a food particle encountering a digestive enzyme and being absorbed across the epithelium and into the blood stream.

Nutrients can cross the intestine wall by either passive or active transport. In passive transport, molecules diffuse into the intestinal cells down a concentration gradient (i.e. they move from a region where they are in high concentration to an area of low concentration.) The sugar xylose enters the blood by passive transport. Active transport requires energy. Amino acids, small peptides, vitamins, and most glucose are moved across the intestine lining by active transport. Once nutrients have moved through the epithelial cells, they are taken up by either capillaries or lacteals and then transported around the body.

Ileum. The main function is absorption of vitamin B12, bile salts and whatever nutrients that were not absorbed by the jejunum. At the point where the ileum joins, the large intestine there is a valve, called the ileocecal valve, which prevents materials flowing back into the small intestine.

Caecum. Is a pouch connected to the large intestine and the ileum. It is separated from the ileum by the ileocecal valve, and is considered to be the beginning of the large intestine. In herbivores the caecum is greatly enlarged and serves as a storage organ that permits bacteria and other microbes time to further digest cellulose. Partially digested food enters the caecum through the ileocecal valve, which is normally closed. The valve occasionally opens to allow food material in. As there is only one opening to the caecum, digesta must move in and out to the caecum through the same opening.

Large intestine. In addition to the caecum the large intestine is made up of the ascending colon, transverse colon, sigmoid colon, rectum, and anus. Much of the large intestine comprises the colon, which is shorter in length but larger in diameter than the small intestine. The colon is involved in the active transport of sodium and absorption of water by osmosis, from the digested material that it contains. It also provides an environment for bacteria to grow and reproduce. These symbiotic bacteria produce important vitamins such as vitamin K, thiamine, and riboflavin, required by the animal for proper growth and health. Finally, the large intestine eliminates wastes. Undigested and unabsorbed food, as well as other body wastes, leave the intestine in the form of faces, via the rectum and anus.

Rumen microorganisms

The rumen provides a suitable environment with substantial food supply for the growth and reproduction of microorganisms. The absence of oxygen (anaerobic) concentration favors the growth of special species of bacteria, those that can digest plant cell walls (cellulose and hemicelluloses) and produce simple sugars (glucose). To grow, microorganisms ferment glucose for energy and produce volatile fatty acids (VFA) as fermentation products. The VFA cross the rumen wall and serve as sources of energy for the ruminant (Figure 6.2). Bacteria can use ammonia or urea as a nitrogen source to produce amino acids. With no bacterial conversion, ammonia, and urea would be useless for ruminants. However, bacterial proteins produced in the rumen are digested in the small intestines and are the main source of amino acids for animal.

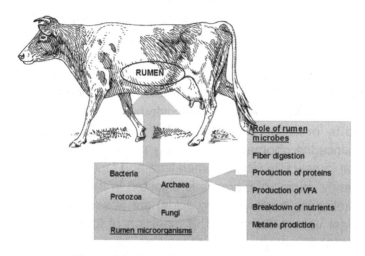

Figure 6.2. Role of rumen microorganisms

Microbes in the reticulorumen include bacteria, protozoa, fungi, archaea, and viruses. Bacteria, along with protozoa, are the predominant microbes and by mass account for 40-60% of total microbial matter in the rumen. They are categorized into several functional groups, such as fibrolytic, amylolytic, and proteolytic types, which preferentially digest structural carbohydrates, non-structural carbohydrates, and protein, respectively. Protozoa (40-60% of microbial mass) derive their nutrients through phagocytosis of other microbes, and degrade and digest feed carbohydrates, especially starch and sugars, and protein. Although protozoa are not essential for rumen functioning, their presence has pronounced effects. Ruminal fungi make up only 5-10% of microbes and are absent on diets poor in fiber. Despite their low numbers, the fungi still occupy an important niche in the rumen because they hydrolyze some ester linkages between lignin and hemicellulose or cellulose, and help break down digesta particles. Rumen Archaea, approximately 3% of total microbes, are mostly autotrophic methanogens and produce methane through anaerobic respiration. Most of the hydrogen produced by bacteria, protozoa and fungi is used by these methanogens to reduce carbon dioxide to methane. The maintenance of low partial pressure of hydrogen by methanogens is essential for proper functioning of the rumen. Viruses are present in unknown numbers and do not contribute to any fermentation or respiration activity. However, they do lyse microbes, releasing their contents for other microbes to assimilate and ferment in a process called microbial recycling, although recycling through the predatory activities of protozoa is quantitatively more important.

Microbes in the reticulorumen eventually flow out into the omasum and the remainder of the alimentary canal. Under normal fermentation conditions the environment in the reticulorumen is weakly acidic and is populated by microbes that are adapted to a pH between roughly 5.5 and 6.5; since the abomasum is strongly acidic (pH 2 to 4), it acts as a barrier that largely kills reticulorumen flora and fauna as they flow into it. Subsequently, microbial biomass is digested in the small intestine and smaller molecules (mainly amino acids) are absorbed and transported in the portal vein to the liver. The digestion of these microbes in the small intestine is a major source of nutrition, as microbes usually supply some 60 to 90% of the total amount of amino acids absorbed. On starch-poor diets, they also provide the predominant source of glucose absorbed from the small intestinal contents. Under conditions of ruminal acidosis, when the environment of the reticulorumen has become too acidic (usually due to excessive fermentation of starches and sugars into VFA and lactate), microbes that favor a lower pH may start to dominate the ecosystem of the reticulorumen. This gives rise to rumen acidosis and often feed intake of the ruminant will drop.

Some important species of rumen bacteria

Fibrobacter succinogenes is the predominant cellulolytic Gram-negative bacterial species in the rumen. It ferments glucose and produces acetate and succinate as waste products. Most ATP is derived from substrate level phosphorylation but some is also derived from anaerobic respiration coupled to the production of succinate.

Ruminococcus flavifaciens is a Gram-positive, cellulolytic bacterium. It is the most active species involved in the digestion of plant cell walls due to its high cellulase and hemicellulase activity. It produces hydrogen, acetate and succinate as end products. The host uses the acetate as an oxidisable substrate and the succinate is used as a growth substrate by some propionate producers. The hydrogen is an important source of reducing equivalents supporting the growth of the methanogens and the homoacetogens.

Megasphaera elsdenii is a Gram-negative coccus and is the predominant bacterial species in the rumen of young ruminants. It is important because it ferments glucose to propionate that is then available to the host for gluconeogenesis.

Selenomonas ruminantium is a non-cellulolytic Gram-negative species that ferments glucose and occurs in large numbers when ruminants are fed grain. This species is an important producer of propionate that is used by the host ruminant for gluconeogenesis.

Veillonella parvula is a Gram-negative bacterium that uses lactate as a growth substrate, from which it makes acetate and propionate. Acetate is produced as the end product of a metabolic pathway that generates ATP by substrate level phosphorylation. Succinate is an intermediate in the propionate production pathway and this organism takes up succinate produced by other bacterial species and uses it as a growth substrate. The use of succinate in this way allows veillonella to make additional ATP by a chemiosmotic mechanism based on dissipation of an electrochemical transmembrane sodium ion gradient. The energy needed to create the gradient comes from the exergonic decarboxylation of methylmalonyl CoA produced from succinate.

Butirivibrio fibrisolvens are Gram-negative cellulolytic bacteria, common in the rumen, producing acetate and butyrate by fermentation of glucose and other substrates.

Lactobacillus ruminis is predominant in the reticulo-rumen of young animals. It is an important glucose fermenter and produces mainly lactate. *Streptococcus bovis* is another important glucose fermenter also producing mainly lactate. Normally lactate is a valuable gluconeogenic substrate for the host ruminant in addition to being used as a growth substrate by propionate producers such as *Veillonella parvula*. These two lactate producers can however create highly acidic conditions in the rumen if the animal eats excessive amounts of readily fermentable carbohydrate. This can lead to the serious clinical condition known as lactic acidosis that is part of the cereal overeating syndrome.

Methanobacterium ruminantium and *Methanosarcina barkeri* are important rumen methanogens. All methanogens are strict anaerobes and they obtain their ATP by a form of anaerobic respiration linked to a chemiosmotic mechanism. They live in syntrophic relations with other bacteria that produce hydrogen and carbon dioxide (as waste products) which the methanogens use for their growth. Uncertainty exists as to whether an electrogenic sodium ion or a proton gradient is generated by the electron transfer processes that use hydrogen to reduce carbon dioxide. Methanogens are very important because their abundance in the reticulo-rumen contributes to determining feed conversion efficiency of the host ruminant.

Acetomaculum ruminis and *Ruminococcus schinkii* are homoacetogens with growth characteristics somewhat similar to the methanogens in that they can grow as chemoautotrophs with CO_2 as their source of carbon for growth and they use hydrogen to supply reducing equivalents for its reduction. The reduced product is acetate. In this growth mode, they derive their ATP by a form of anaerobic respiration that depends on electron transport maintaining a transmembrane electrogenic sodium ion gradient that provides the free energy to drive a membrane bound ATP synthase.

As indicated previously these microorganisms are very adaptable. They are not restricted to growing as chemoautotrophs. They can also grow as chemoheterotrophs fermenting glucose to derive some of their ATP by substrate level phosphorylation. When they do this, they produce carbon dioxide and hydrogen and then they can obtain an extra yield of ATP by anaerobic respiration using the hydrogen to reduce the carbon dioxide to acetate as in the autotrophic growth mode.

Protozoa

The second most common microbe (by mass) in the bovine rumen are ciliated protozoa. The most common protozoans in the rumen are of the geniuses, Epidinium, Entodinium, Diplodinium, and Holotrich ciliates, which are not actually a genus but a type of protozoa defined by its hairy covering of cilia. Specific examples of common protozoa include Ophryoscolex monoacanthus, Entodinium exiguum, Eudiplodinium maggii, and Isotricha intestinalis. The composition of protozoa is correlated with the diet of the animal; different dietary intakes greatly affect the amount and composition of the ruminal protozoa. Starvation tends to decrease all protozoal counts, although in some cases Holotrichs are known to actually flourish under those conditions. Dietary additives like tylosin increases the population of protozoa in the rumen. Some protozoa end up going further down the digestive tract, while others sequester themselves into the rumen. Much like the archaea and bacteria in the rumen, the protozoa play a wide range of important roles in their niche.

The big role the protozoans play in the bovine rumen is to digest material that the animal cannot normally do on its own. They help the animal to metabolize plant material, lipids, and proteins. Protozoans can synthesize long chain fatty acids by using small precursor molecules found in the ruminal fluid. It is thought that phospholipids are synthesized from exogenous precursors in protozoans by incorporating linoleic acids with sterol esters. Ciliated protozoa in the rumen are thought to be responsible for roughly 35% of the digestion of plant material in the rumen. Different types of protozoa digest different parts of plant material; large protozoa prefer to ingest and degrade plant structural polymers, whereas small protozoa utilize storage polymers and sugars. When protozoa such as Epidinium are colonized on plant material, a large amount of damage to the plant material is seen. Ciliates also attach to plant material in the rumen and are able to find them through chemotaxis to certain sugars using a specific organelle on their dorsolateral surface. These ciliates after attaching can then break down storage sugars, hemicellulose, pectin, and to a lesser degree, cellulose. The ability of protozoa to metabolize and assimilate protein is also key to the animal's well-being, and it does this by breaking down dietary and microbial protein. This is especially important because the protozoans are thought to be able to synthesize many amino acids that the diets of the host usually lack. On the flip side to this, protozoa can actually compete detrimentally with the host animal for protein in the rumen; this is especially true in starving animals and negligible in well fed animals.

Protozoa in the rumen also have a partially symbiotic relationship to their bacterial neighbors. Although protists in the rumens all actively ingest and feed on bacteria, the digestive waste products are released into the rumen and can be taken up by the microflora). Protozoans also help facilitate methanogenesis by taking up oxygen that creates a more anaerobic environment which allows the anaerobic bacteria and archaea to carry out methanogenesis. It is thought that they do this by removing oxygen from the liquid parts of the rumen and then moving to the reticulum of the rumen, which creates an environment in the rumen where methanogenesis can take place.

Protozoans themselves also serve as a source of nutrition for the hosts, as they themselves are often digested by the animal's digestive system. Compared to bacteria, rumen protozoa contain a high percentage of unsaturated fatty acids and this is thought to be an important source of lipids for the animal hosts. Up to 27% of the lipids in the rumen digesta is thought to be from Holotrich protists. Ciliated protozoa also protect unsaturated fatty acids from being fully saturated in the rumen by incorporating these fatty acids into their membrane phospholipids. Protozoans are also a big source of protein for the host animals, and in cows microbial protozoa supply around 20% of the total protein needed. Protozoal protein shares the same composition as bacterial protein in the bovine rumen; however, protozoal protein is more easily digested. A large percentage (~50%) of these proteins are made of the following four amino acids: glutamic acid, leucine, lysine and isoleucine.

Anaerobic rumen fungi

Anaerobic fungi are the significant constituent of rumen microbiota in livestock that rely on poor-quality fibrous diets. Such fungi colonize plant fragments in the rumen of cattle and other herbivores. Through rhizoidal growth, they penetrate the plant cell wall and increase the area susceptible to enzymatic attack. The fibrolytic enzymes produced perform in concert to efficiently degrade cellulose and hemicellulose to simple sugars with the end-products acetate, lactate, ethanol, formate, CO_2 and H. In contrast to the other rumen microbes, fungal enzymes also break ester linkages between lignin and hemicelluloses.

Therefore, these fungi are found attached mainly with the lignified tissues that remain in the rumen for extended periods and maintain highest number in animals receiving high-fiber diets but lack in rumen of animals receiving leafy forage, due to the shorter retention period of such feedstuffs. These qualities of anaerobic fungi together with the degree of colonization and growth on fibrous plant fragments collectively suggest that manipulation of such a group of microbes has immense potential for boosting digestive performance in the rumen and ultimately higher animal production response.

Rumen archaea

When these microscopic organisms were first discovered, they were considered bacteria. However, when their ribosomal RNA was sequenced, it became obvious that they bore no close relationship to the bacteria and were, in fact, more closely related to the eukaryotes. For a time, they were referred to as archaebacteria, but now to emphasize their distinctness, and then were called archaea. Based on the analysis of the global data set, the majority (92.3%) of rumen archaea detected in total rumen contents can be placed in three genus-level groups. These are *Methanobrevibacter* (61.6%), *Methanomicrobium* (14.9%), and a large group of uncultured rumen archaea labeled here as rumen cluster C, or RCC (15.8%). *Methanobrevibacter* spp. have been considered the dominant methanogens in the rumen.

Within the genus *Methanobrevibacter*, the cloned sequences fall into two major clades. One clade, defined by the species *Methanobrevibacter gottschalkii*, *Methanobrevibacter thaueri*, and *M. millerae*, contains the larger part of the *Methanobrevibacter*-related clones (a mean of 33.6% of rumen archaea). This group is designated the *M. gottschalkii* clade. The other major clade, defined by M. ruminantium and *M. olleyae*, contains 27.3% of rumen archaea, and is designated the M. ruminantium clade. Members of these two clades were found in nearly all of the data sets. *Methanobrevibacter* spp also appear to be early colonizers of the developing rumen. Members of other *Methanobrevibacter* spp, including *M. smithii* and *Methanobrevibacter wolinii*, appear to be rare.

Rumen pH and its regulation

The pH values in the rumen are variable, and influenced by factors such as the type of food and time of consumption. Normal physiological pH values are between 5.4 and 6.9. Three aspects must be taken into account as regulatory factors:

1. Influence of the VFA in the increase in acidity. During fermentation, processes VFA are produced. Increased in acidity is reached after about 3 hours of ingestion, being generally higher when fermentation processes are more intense.
2. The amount of saliva secreted during mastication and rumination. Since saliva has a pH between 8.1 and 8.3, this makes a fatty acid neutralizing agent, property conferred by the salts (bicarbonate and phosphate of sodium and potassium). The amount of saliva secreted fluctuates between 100 and 180 liters in the bovine.
3. The rate of absorption of VFA functions as a buffer acidity when they are produced. A lesser is degree of dissociation, the greater is the speed. Lowering the pH, the degree of dissociation is reduced and therefore, increasing the absorption rate and a certain stabilization of the pH is achieved.

Rumination process is important in the dynamic regulation of pH due to great contribution of saliva to the environment. During this process, three times more saliva is produced than during the mastication. The type of food and its physical and chemical properties affects rumination time.

Digestibility of structural polysaccharides (cell wall)

The main structural polysaccharides of the cell wall of forages include hemicellulose, cellulose and peptic compounds. Peptic substances are rare in the grasses, but are found in large amounts in legumes, sometimes comprising a little more of 100 g kg^{-1} of dry matter (DM). Peptic substances are found in high concentrations in the middle lamella, especially in dicotyledonous plants. However, low concentrations of peptic compounds are found in the primary cell wall. No deposition occurs during thickening pectin secondary cell wall.

Hemicelluloses represent a very diverse group of polysaccharides including the arabinoxylans, xyloglucans, gluconomananas and mixed bonds of glycans. The hemicellulolytic arabinoxylans are the predominant polysaccharide in the primary cell walls of grasses, while xyloglucans hemicellulose are the most abundant in the primary cell wall of legumes. The glucurono-arabinoxylans, hemicellulolytic polysaccharides are the mainly deposited in the secondary thickening of the cell walls of grasses and legumes. Hemicelluloses usually contribute a larger proportion to the total polysaccharides in the cell walls in primary secondary cell walls. Forage hemicellulose are less digested than pectins by ruminants. The degree of digestion of the hemicelluloses varies considerably and is dependent on factors such as genotype, maturity and feeding practices.

Cellulose is the most abundant in primary and secondary walls of forages polysaccharide. The cellulose synthesized during the thickening of the cell walls is considerably crystalline cellulose from primary cell walls. As hemicelluloses, cellulose forage is not digested completely.

Leaves vs stems

The leaves and stems of grasses differ significantly in their chemical composition and digestibility. Leaves generally contain more protein and low concentration of structural polysaccharides and lignin that stems. Digestibility *in vitro* and *in situ* of leaves, usually are considerably higher than that of the stems. The increased digestibility observed in the leaves is because the leaves contain less cell wall and the cell wall polysaccharides are more digestible. The magnitude of the differences in digestibility of leaves and stems is positively correlated with the increase in the level of maturity of the grasses. However, in the early stages of growth of grasses, the *in vitro* digestibility of the stems may be slightly higher than that of the leaves.

Factors affecting the degree of digestion of the cell wall

The increase in the physiological maturity of forage plants causes a tremendous reduction in the leaf:stem ratio. In grasses, the content of CW and lignin increase with increasing maturity. Consequently, the leaf and stem digestibility decline as maturity increased. At high rates of intakes in ruminants, is associated with a reduction of the dry matter and cell wall polysaccharides. A high consumption, the decreases in cell wall digestibility is two to three times higher than non-structural carbohydrates. The decrease in digestibility is because there is a lesser degree of digestion because there is less residence time of food in the rumen and throughout the digestive tract.

Associative effects. Can be defined as the interaction between nutrients in different ingredients in a ration which result in performance that is greater or less than expected from the individual ingredients. The feeding value of forage mixtures from permanent and temporary multi-species grasslands cannot always be precisely defined. Indeed, the digestibility and feed intake of a combination of forages can differ from the balanced median values calculated from forages considered separately. Interactions between forages can lead to associative effects on intake and digestion in three main situations:

(i)   Increased intake that can be observed with grass and legume association can be explained by fast digestion of the soluble fraction of legumes, and a higher rate of particle breakdown and passage through the rumen.
(ii) Increased digestion, when a poor forage is supplemented by a high nitrogen content source, might be explained by stimulation of the microbial activity.
(iii) Modification of digestive processes in the rumen, including proteolysis and methane production when are present certain bioactive secondary metabolites such as tannins, saponins or polyphenol oxidase.

According to the type and concentration of these compounds in the diet, the effects can be favorable or unfavorable on intake and digestive parameters. Reports of associative effects between forages show a large variability among studies. This reflects the complexity and multiplicity of nutritional situations affecting intake and the rumen function in a given animal. In order to provide more reliable information, further accumulation of data combining *in vitro* and *in vivo* studies is required. A better understanding of the associative effects between forages could help to optimize feed use efficiency, resulting in greater productivity, a reduction of the environmental impact of animal emissions and more sustainable animal production. In addition, the associative effects could be largely avoided if the rumen pH is maintained above that level inhibitory to cellulolysis. is suggested that this could be achieved by offering roughage portion of the diet either long or chopped, so as to stimulate rumination and salivation, by offering the concentrate in such a way as to minimize the risk of depressing material, such as bicarbonate salts, in the diets.

Neutral detergent fiber and dry matter digestibilities and metabolizable energy of cultivated grasses

In situ neutral detergent fiber digestibility

The effective degradability of neutral detergent fiber (EDNDF) varied among cultivated grasses (Table 6.3). *Cenchrus ciliaris,* harvested in Marin, NL, Mexico in 1994, had highest annual mean value, but, in that same year, *D. annulatum* collected the same region, had the lowest annual mean. There were also differences between seasons, apparently during wet seasons (summer and fall); the EDNDF was higher compared to that of the dry season (winter and spring). As previously mentioned, with increasing maturity of the plant, during winter and spring, lignin content also increases which is considered the main cause of the reduction in digestion of the cell wall.

Table 6.3. Seasonal variation of the effective degradability of neutral detergent fiber (EDNDF; % dry matter) and the characteristics of the *in situ* digestibility of cultivated grasses collected in the counties and dates of the state of Nuevo León, Mexico

Grasses	Places and dates of collection	Concept	Winter	Spring	Summer	Fall	Annual mean
Cenchrus ciliaris, Common	Marin, N.L., Mexico (1994)	EDNDF	52	59	60	74	61
		a, %	36	41	44	59	45
		b, %	27	35	23	23	27
		c, % h^{-1}	6	6	6	6	6
Cenchrus ciliaris, Common	Marin, N.L., Mexico (1998)	EDNDF	35	32	58	43	42
		a, %	16	15	16	8	14
		b, %	30	25	75	58	47
		c, % h^{-1}	5	5	3	5	5
Cenchrus ciliaris, Common	Teran, N.L., Mexico (2001-02)	EDNDF	39	47	53	50	47
		a, %	19	26	32	23	25
		b, %	29	28	31	36	31
		c, % h^{-1}	9	10	6	9	9
Cenchrus ciliaris, Common	Linares, N.L., Mexico (1998-99)	EDNDF	28	43	39	40	38
		a, %	9	8	23	4	11
		b, %	41	49	24	58	43
		c, % h^{-1}	2	6	6	4	4
Cynodon dactylon	Linares, N.L., Mexico (1998-99)	EDNDF	30	39	36	40	36
		a, %	9	18	15	6	12
		b, %	35	37	29	66	42
		c, % h^{-1}	4	3	7	3	4
Cynodon dactylon	Marin, N.L., Mexico (1994)	EDNDF	34	36	39	40	37
		a, %	13	18	22	16	17
		b, %	35	31	42	41	37
		c, % h^{-1}	6	6	6	6	6
Cynodon dactylon II	Marin, N.L., Mexico (1994)	EDNDF	46	50	47	53	49
		a, %	21	21	27	23	23
		b, %	43	53	44	53	48
		c, % h^{-1}	6	5	6	6	6
Dichanthium annulatum	Marin, N.L., Mexico (1994)	EDNDF	29	33	39	31	33
		a, %	13	15	17	13	15
		b, %	27	30	39	32	32
		c, % h^{-1}	6	6	6	6	6
Dichanthium annulatum	Linares, N.L., Mexico (1998-99)	DEFDN	32	46	37	48	41
		a, %	11	8	11	6	9
		b, %	33	57	38	60	47

		c, % h⁻¹	5	5	6	6	5
Panicum coloratum	Linares, N.L., Mexico (1998-99)	EDNDF	33	32	30	37	33
		a, %	12	6	12	7	10
		b, %	40	43	30	55	42
		c, % h⁻¹	3	4	4	3	3
Rhynchelytrum repens	Teran, N.L., Mexico (2001-02)	EDNDF	38	39	62	43	46
		a, %	18	25	32	23	24
		b, %	27	19	44	28	29
		c, % h⁻¹	9	9	6	9	8

EDNDF = calculated with a ruminal rate of passage of 2 % h⁻¹
a = is an intercept representing the portion of the NDF solubilized at the beginning of digestion (t = 0): b = is the portion of NDF slowly degraded in the rumen; c = is the constant rate of disappearance of the fraction b;
Obtained from: Ramirez *et al.* (2002); Ramirez *et al.* (2003); Ramirez (2003); Ramirez *et al.* (2004); Ramirez *et al.* (2005);

An external factor also limits the cell wall degradability is present on the cuticular surface wall of the plant. The cuticle is indigestible for microorganisms and therefore serves as a barrier preventing access of microorganisms to the outer surface of the plant. The cuticle is increased under high temperatures, light and acidity. It is found in large concentrations in the axial base of the leaves.

Dry matter digestibility and metabolizable energy

It has been found that the acid detergent fiber (ADF) is a better indicator for determining the nutritional value compared to crude fiber, because ADF contain cellulose and lignin, and the dry matter digestibility (DMD) decreased with increasing lignin. It was also investigated that quality of five forage species and stated that N content and ADF as two important factors in determining the metabolizable energy requirements of livestock. Thus, the data of DMD and metabolizable energy (ME) were obtained from the following predicted equations:

DMD (DMD, % = 83.5 - 0.824ADF% + 2.626N%)
Digestible energy (DE, Mcal kg⁻¹ = 0.27 + 0.0428DMD%)
Metabolizable energy (ME. Mcal kg⁻¹ = 0.821DE Mcal kg⁻¹).

Most cultivated grasses cut in some counties of the state of Nuevo Leon, Mexico (Figure 6.3) had DMD percentages below 50%, except for *Cenchrus ciliaris* cut in Linares county of the state of Nuevo Leon,

Mexico, *Cynodon dactylon* Cross II, *Dichanthium annulatum* and *Panicum coloratum* that reached 50% or more. During spring, most grasses had the highest DMD and in autumn were lowest.

The knowledge of how the animal uses the ME for its metabolic functions is extremely important, since this efficiency varies according to the physiological function (maintenance, gain, pregnancy etc), and to the concentration of ME in the diet. The ME of all grasses (Figure 6.4) followed the same tendency as DMD. It seems that most grasses had ME values compared to hay sun-cured full bloom (2.00 Mcal kg-1) and alfalfa fresh full bloom (1.81 Mcal kg-1).

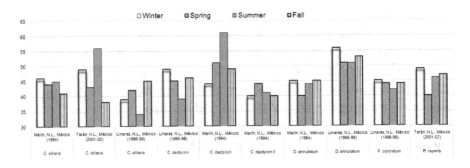

Figure 6.3. Dry matter digestibility (%, dry matter) by cultivated grasses cut in different counties of the state of Nuevo Leon, Mexico

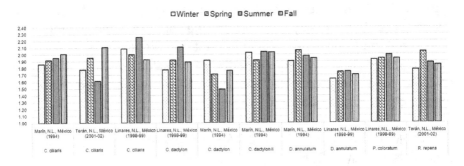

Figure 6.4. Metabolizable energy (Mcal kg^{-1}, dry matter) by cultivated grasses cut in different counties of the state of Nuevo Leon, Mexico

The addition of 75 kg ha^{-1} of N from urea increased the dry matter intake of cattle eating grass, but the digestibility of dry matter and cell wall were not affected by N fertilization. Similar responses in EDNDF (Table 6.3), DMD (Figure 6.5) and ME (Figure 6.6) have been reported in new genotypes of buffelgrass (*Cenchrus ciliaris*) fertilized with N from urea at 120 kg ha^{-1} and planted in Teran county of the state of Nuevo Leon, Mexico.

Table 6.4. Seasonal variation of the effective degradability of neutral detergent fiber (EDNDF; % dry matter) and the characteristics of the *in situ* digestibility of the hybrid buffelgrass Nueces and five new genotypes of buffelgrass collected in different dates in Teran county of the state of Nuevo Leon, Mexico

Genotypes	Concept	Dates of collection			
		Aug. 1999	Nov. 1999	Nov. 2000	Jun. 2000 (Fertilized)
Cenchrus ciliaris Nueces	EDNDF	41	56	70	52
	a, %	27	34	49	20
	b, %	30	31	34	46
	c, % h^{-1}	5	6	4	6
Cenchrus ciliaris 307622	EDNDF	40	64	65	42
	a, %	16	36	40	17
	b, %	43	36	41	38
	c, % h^{-1}	5	5	4	5
Cenchrus ciliaris 409252	EDNDF	39	59	58	37
	a, %	10	24	37	16
	b, %	48	33	38	32
	c, % h^{-1}	4	6	3	6
Cenchrus ciliaris 409375	EDNDF	42	64	64	44
	a, %	16	25	44	19
	b, %	45	35	33	35
	c, % h^{-1}	5	6	4	6
Cenchrus ciliaris 409460	EDNDF	38	55	66	44
	a, %	15	33	44	20
	b, %	42	40	43	36
	c, % h^{-1}	4	4	3	6
Cenchrus ciliaris 443	EDNDF	47	65	55	43
	a, %	9	30	31	18
	b, %	44	35	40	36
	c, % h^{-1}	5	6	4	6

EDNDF = calculated with a ruminal rate of passage of 2 % h^{-1}; a = is an intercept representing the portion of the NDF solubilized at the beginning of digestion (t = 0); b = is the portion of NDF slowly degraded in the rumen; c = is the constant rate of disappearance of the fraction b. Obtained from: García-Dessommes *et al.* (2003ab)

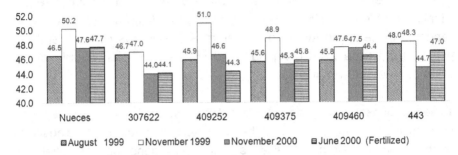

Figure 6.5. Dry matter digestibility (%, dry matter) of the hybrid buffelgrass Nueces and five new genotypes of *Cenchrus ciliaris* collected in different dates in Teran county of the state of Nuevo Leon, Mexico

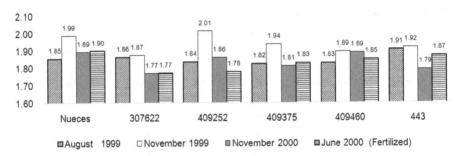

Figure 6.6. Metabolizable energy (Mcal kg-1, dry matter) of the hybrid buffelgrass Nueces and five new genotypes of *Cenchrus ciliaris* collected in different dates in Teran county of the state of Nuevo Leon, Mexico

The most important response in the increasing the maturity, of most forages, is the tremendous reduction of the leaf:stem ratio. In grasses, leaves and stems increase their cell wall and lignin content as maturity increases. Consequently, the digestibility of the leaves and stems of grasses decline as forage maturity increases, but the rate of decrease is greater in the stems than in the leaves. Similar response was observed in cultivated grasses shown in Table 6.5. It was detected that the EDNDF was greater in leaves than in the stems and EDNDF was lower in winter when the grasses are mature (Table 6.5).

Table 6.5. Seasonal variation of the effective degradability of neutral detergent fiber (EDNDF; % dry matter) and the characteristics of the *in situ* digestibility of stems and leaves of four cultivates grasses collected in different dates in Teran county of the state of Nuevo Leon, Mexico

Grasses	Places and dates of collection	Parts	Concept	Spring	Primavera	Summer	Fall	Annual mean
Cenchrus ciliaris Common	Linares, N.L., México (1998-99)	Leaves	EDNDF	45	57	49	52	51
			a, %	12	17	25	5	15
			b, %	65	64	48	74	63
			c, % h⁻¹	3	4	3	5	4
		Stems	EDNDF	19	38	32	37	31
			a, %	8	10	17	6	10
			b, %	18	42	20	45	31
			c, % h⁻¹	4	5	8	7	6
Cynodon dactylon	Linares, N.L., México (1998-99)	Leaves	EDNDF	37	54	38	34	41
			a, %	13	21	15	6	14
			b, %	38	68	33	51	47
			c, % h⁻¹	5	2	6	3	4
		Stems	EDNDF	19	36	35	34	31
			a, %	5	7	12	5	7
			b, %	23	56	34	49	41
			c, % h⁻¹	4	3	6	3	4
Dichanthium annulatum	Linares, N.L., México (1998-99)	Leaves	EDNDF	42	56	49	54	50
			a, %	12	5	17	11	11
			b, %	45	81	48	60	58
			c, % h⁻¹	5	5	5	3	4
		Stems	EDNDF	22	56	30	34	35
			a, %	8	13	12	4	9
			b, %	22	63	30	52	41
			c, % h⁻¹	5	5	6	7	6
Panicum coloratum	Linares, N.L., México (1998-99)	Leaves	EDNDF	43	41	36	38	39
			a, %	13	18	15	4	12
			b, %	62	48	34	60	51
			c, % h⁻¹	2	2	4	3	3
		Stems	EDNDF	30	33	26	33	30
			a, %	11	11	11	7	10
			b, %	34	59	21	52	42
			c, % h⁻¹	3	2	6	3	3

EDNDF = calculated with a ruminal rate of passage of 2 % h⁻¹; a = is an intercept representing the portion of the NDF solubilized at the beginning of digestion (t = 0); b = is the portion of NDF slowly degraded in the rumen; c = is the constant rate of disappearance of the fraction b.
Data obtained from Ramirez *et al.* (2001ab); Foroughbackhch *et al.* (2001); Ramirez *et al.* (2003); Ramirez (2003); Ramirez *et al.* (2005).

The DMD of leaves (Figure 6.5) and stems (Figure 6.6), and ME of leaves (Figure 6.7) and stems (Figure 6.8) in the four cultivated grasses, followed the same pattern as EDNDF. Leaves had higher values than stems and were lower in winter when the grasses were mature. It has been argued that the single most important determinant of forage quality is stage of maturity of the plant when harvested. Grasses, like animals, grow and mature over time. The mature grass is one that has developed reproductive components to the point of generating seeds. Immature grass is the lush rapidly growing plant prior to reproductive parts (seeds, flowers) development. Mature grasses contain greater amounts of cell wall structural components, as measured by neutral detergent fiber, and lignin for cell wall reinforcement, as reflected in acid detergent fiber amounts. These cell wall components allow the large mature grass to stand upright, rather than to fall over under its own weight. Grass cells, in contrast to animal cells, have a rigid cell wall. This increase in lignin and fiber results in a dilution of energy, protein and other nutrients as well as a decline in nutrient digestibility.

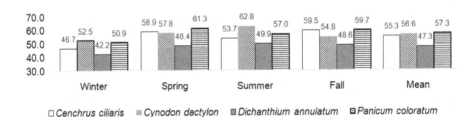

Figure 6.7. Dry matter digestibility (%) of leaves

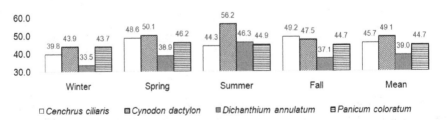

Figure 6.8. Dry matter digestibility (%) of stems

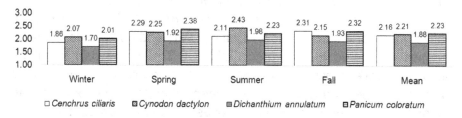

Figure 6.9. Metabolizable energy (Mcal kg-1) of leaves

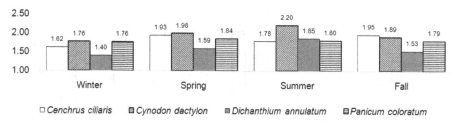

Figure 6.10. Metabolizable energy (Mcal kg-1) of stems

Neutral detergent fiber and dry matter digestibilities and metabolizable energy of native grasses

The EDNDF of native grasses collected in Marin county of the state of Nuevo Leon, Mexico was different between grasses and seasons (Tables 6.6). The NDF of the grasses *H. belangeri and P. hallii* was more degraded by the microorganisms in the rumen than the NDF of other grasses. *Aristida longiseta* and *B. gracilis* had the lowest EDNDF. In general, during the winter native grasses had the lowest EDNDF. Similar trend as EDNDF was observed in DMD (Figure 6.9) and ME (Figure 6.10) in native grasses, and lower DMD and ME were registered in winter when the grasses were mature.

Native grasses such as *B. trifida* and *B. curtipendula* had the lowest EDNDF values compared with other native grasses that grow in Teran county of the state of Nuevo Leon, Mexico. *Panicum obtusum* resulted with the highest value (Table 6.7). In general, during the wet season (summer and fall) most grasses had the highest EDNDF compared to the dry season (winter and spring). A similar trend as EDNDF was found in the DMD (Table 6.8) and ME (Table 6.9) of native grasses. In general, low EDNDF, DMD and ME values of native grasses compared to cultivated (Tables 6.3, 6.4 and 6.5) might be because native grasses contained more amounts of hemicellulose than cultivated grasses. It has been reported that xylose, an important component of hemicelluloses, is always present in large concentrations in less degradable plant tissues.

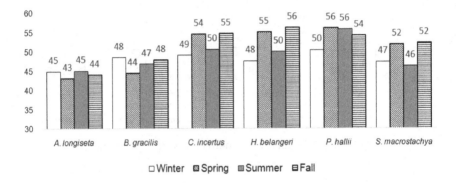

Figure 6.10. Dry matter digestibility (%) of native grasses

Table 6.6. Seasonal tendency of the effective degradability of neutral detergent fiber (EDNDF; % dry matter) and the characteristics of the in situ digestibility of native grasses collected in Marin county of the state Nuevo L., Mexico 1994

| Native grasses | Concept | Seasons | | | | Annual |
		Winter	Spring	Summer	Fall	mean
Aristida longiseta	EDNDF	37	42	45	56	45
	a, %	25	29	32	43	32
	b, %	22	20	20	21	21
	c, % h^{-1}	6	6	6	5	6
Bouteloua gracilis	EDNDF	54	57	54	60	56
	a, %	37	43	40	35	39
	b, %	28	25	24	39	29
	c, % h^{-1}	6	6	5	5	6
Cenchrus incertus	EDNDF	50	54	49	65	55
	a, %	24	26	22	52	31
	b, %	44	45	47	20	39
	c, % h^{-1}	6	6	6	5	6
Hilaria belangeri	EDNDF	40	53	61	65	55

	a, %	40	39	42	49	43
	b, %	21	33	33	25	28
	c, % h⁻¹	6	5	5	5	5
Panicum hallii	EDNDF	54	60	59	76	62
	a, %	37	35	36	38	37
	b, %	30	26	25	37	30
	c, % h⁻¹	5	4	4	4	4
Setaria macrostachya	EDNDF	49	63	60	69	60
	a, %	42	58	35	44	45
	b, %	29	18	23	33	26
	c, % h⁻¹	5	4	4	4	4

EDNDF = calculated with a ruminal rate of passage of 2 % h⁻¹; a = is an intercept representing the portion of the NDF solubilized at the beginning of digestion (t = 0): b = is the portion of NDF slowly degraded in the rumen; c = is the constant rate of disappearance of the fraction b; Obtained from: Ramirez *et al.* (2004)

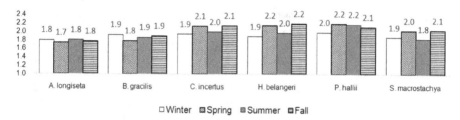

Figure 6.10. Metabolizable energy (Mcal kg⁻¹, dry matter) of native grasses

Table 6.7. Seasonal tendency of the effective degradability of neutral detergent fiber (EDNDF; % dry matter) and the characteristics of the in situ digestibility of native grasses collected in Teran county of the state of Nuevo Leon, México in 2001 y 2002

Native grasses	Concept	Seasons				Annual
		Winter	Spring	Summer	Fall	mean
Bouteloua curtipendula	EDNDF	35	38	36	38	37
	a, %	16	19	18	20	18
	b, %	27	24	26	24	25
	c, % h⁻¹	5	4	4	4	4
Bouteloua trifida	EDNDF	33	37	38	39	37
	a, %	16	21	23	21	20
	b, %	24	21	20	25	22
	c, % h⁻¹	5	4	4	4	4
Brachiaria fasciculata	EDNDF	38	36	58	45	44
	a, %	18	19	36	21	23

	b, %	28	23	32	32	29
	c, % h^{-1}	5	4	4	4	4
Digitaria insulares	EDNDF	39	40	50	40	42
	a, %	22	21	32	25	25
	b, %	23	24	25	19	23
	c, % h^{-1}	5	4	4	4	4
Chloris ciliata	EDNDF	38	44	48	46	44
	a, %	26	23	26	23	24
	b, %	18	28	29	32	27
	c, % h^{-1}	6	5	5	5	5
Leptochloa filiformis	EDNDF	38	40	44	49	43
	a, %	18	19	20	23	20
	b, %	27	29	34	34	31
	c, % h^{-1}	6	5	5	5	5
Panicum hallii	EDNDF	36	39	49	45	43
	a, %	18	21	26	24	22
	b, %	21	24	34	29	27
	c, % h^{-1}	9	8	7	9	8
Panicum obtusum	EDNDF	35	51	59	49	49
	a, %	20	35	31	26	28
	b, %	20	20	45	31	29
	c, % h^{-1}	8	10	5	8	8
Panicum unispicatum	EDNDF	35	36	43	47	40
	a, %	16	21	26	26	22
	b, %	27	20	24	29	25
	c, % h^{-1}	6	5	5	5	5
Setaria grisebachii	EDNDF	38	41	41	39	40
	a, %	19	25	25	23	23
	b, %	25	21	19	23	22
	c, % h^{-1}	6	5	5	5	5
Setaria macrostachya	EDNDF	33	37	54	40	41
	a, %	18	21	27	22	22
	b, %	21	22	40	24	27
	c, % h^{-1}	6	5	4	5	5
Tridens eragrostoides	EDNDF	41	41	44	38	41
	a, %	24	23	24	19	22
	b, %	23	23	32	25	26
	c, % h^{-1}	6	5	4	5	5
Tridens muticus	EDNDF	34	36	51	37	39
	a, %	16	21	25	21	21
	b, %	24	20	36	22	26
	c, % h^{-1}	6	5	5	4	5

EDNDF = calculated with a ruminal rate of passage of 2 % h-1
a = is an intercept representing the portion of the NDF solubilized at the beginning of digestion (t = 0): b = is the portion of NDF slowly degraded in the rumen; c = is the constant rate of disappearance of the fraction b;
Obtained from: Cobio-Nagao (2004)

Table 6.8. Dry matter digestibility (%) of native grasses collected in Teran county of the state of Nuevo Leon, Mexico

| Native grasses | Seasons | | | | Annual |
	Winter	Spring	Summer	Fall	mean
Bouteloua curtipendula	47.7	47.5	48.9	44.3	47.1
Bouteloua trifida	45.0	48.3	47.7	46.6	47.1
Brachiaria fasciculata	46.3	51.1	55.4	54.9	51.7
Digitaria insularis	55.4	54.8	55.7	52.4	54.6
Chloris ciliata	48.9	49.4	46.6	45.2	47.6
Leptochloa filiformis	48.8	46.9	48.9	51.1	48.9
Panicum hallii	49.5	50.8	53.7	53.1	51.6
Panicum obtusum	47.3	50.7	51.3	51.0	51.7
Panicum unispicatum	53.4	53.4	51.6	51.6	52.1
Setaria grisebachii	46.0	49.1	53.1	47.0	50.0
Setaria macrostachya	48.8	51.5	52.8	45.6	49.8
Tridens eragrostoides	50.7	49.8	52.2	48.0	50.0
Tridens muticus	45.0	44.2	45.4	40.7	43.0

Table 6.9. Metabolizable energy (Mcal kg^{-1}, dry matter) of native grasses collected in Teran county of the state of Nuevo Leon, Mexico

| Native grasses | Seasons | | | | Annual |
	Winter	Spring	Summer	Fall	mean
Bouteloua curtipendula	1.90	1.89	1.94	1.78	1.88
Bouteloua trifida	1.80	1.92	1.90	1.86	1.88
Brachiaria fasciculata	1.85	2.02	2.17	2.15	2.04
Digitaria insularis	2.17	2.15	2.18	2.06	2.14
Chloris ciliata	1.94	1.96	1.86	1.81	1.89
Leptochloa filiformis	1.94	1.87	1.94	2.02	1.94
Panicum hallii	1.96	2.01	2.11	2.09	2.03
Panicum obtusum	1.88	2.00	2.02	2.01	2.04
Panicum unispicatum	2.10	2.10	2.04	2.04	2.05
Setaria grisebachii	1.84	1.95	2.09	1.87	1.98
Setaria macrostachya	1.94	2.03	2.08	1.82	1.97
Tridens eragrostoides	2.00	1.97	2.05	1.91	1.98
Tridens muticus	1.80	1.78	1.82	1.65	1.73

Degradability of protein in grasses

Nutrition of protein in ruminants is a complex process that is greatly impacted by events occurring in the fore stomach and intestine of the ruminant. Dietary crude protein and non-protein N (NPN) are degraded in the rumen and converted into ammonia, which could be used by rumen microorganisms as a precursor for the biosynthesis of amino acids and microbial protein. The microbial protein is subsequently digested in the small intestine of the host and serves as an important source of amino acids that might be used for protein synthesis and converted in animal products. Therefore, the degree of degradation of a particular ingredient or diet, may be a limiting factor in the process of amino acid supplementation to the host and consequently to animal productivity.

One of the factors that most influence on the degradability of the protein in the rumen is the degree of maturity of the forage. It has been found that as the maturity of grasses progresses, usually results in a decrease in voluntary intake and the degree of digestion of protein in the rumen. Table 6.8 shows that the effective degradability of crude protein (EDCP) of grasses grown in the northeastern region of Mexico was different between species and between seasons. In winter, the dry season, EDCP was lower in all grasses. *Cynodon dactylon* grass Cross II had the highest percentage and *D. annulatum* had the lowest annual mean, both grasses growing in Marin county of the state of Nuevo Leon, Mexico. Because *Cynodon dactylon* had a higher EDCP value, places it as a good digestibility grass.

The hybrid buffelgrass Nueces had the highest EDCP in all seasons than the five genotypes of buffelgrass listed in Table 6.9. It seems that fertilization resulted in lower values of EDCP in the hybrid and in the new genotypes of buffelgrass compared with the same plants but unfertilized. Similar response has been reported in studies that found that N fertilization only increases the forage production, but not the in situ digestibility. Thus, N fertilization may not lead to better individual animal performance to the animals consuming these grasses. There are several factors affecting plant response to fertilization such as soil type, soil fertility, soil temperature, and the amount of precipitation. Length of grazing period and date and rate of application can alter effectiveness of nitrogen fertilizer.

Table 6.10, Seasonal variation of the effective degradability of crude protein (EDCP; % dry matter) and the characteristics of the *in situ* digestibility of cultivated grasses collected in the counties and dates of the state of Nuevo León, Mexico

Grasses	Place and date of collection	Concept	Seasons				Annual mean
			Winter	Spring	Summer	Fall	
Cenchrus ciliaris, Common	Marin, N.L., Mexico (1994)	EDCP	59	74	61	52	61
		a, %	36	41	44	59	45
		b, %	27	35	23	23	27
		c, % h^{-1}	8	7	6	6	7
Cenchrus ciliaris, Common	Marin, N.L., Mexico (1998)	EDCP	51	67	75	71	66
		a, %	37	50	55	51	48
		b, %	22	28	28	27	26
		c, % h^{-1}	5	7	4	6	5
Cenchrus ciliaris, Common	Teran, N.L., Mexico (2001-02)	EDCP	52	70	72	75	67
		a, %	49	56	60	49	54
		b, %	19	18	18	18	18
		c, % h^{-1}	4	4	4	4	4
Cenchrus ciliaris, Common	Linares, N.L., Mexico (1998-99)	EDCP	54	74	65	73	67
		a, %	36	52	57	49	49
		b, %	28	33	12	34	27
		c, % h^{-1}	4	5	7	5	5
Cynodon dactylon	Linares, N.L., Mexico (1998-99)	EDCP	56	67	64	69	64
		a, %	35	53	48	53	47
		b, %	37	17	28	29	28
		c, % h^{-1}	4	4	4	3	4
Cynodon dactylon	Marin, N.L., Mexico (1994)	EDCP	65	58	63	77	66
		a, %	48	37	44	62	48
		b, %	26	34	34	23	30
		c, % h^{-1}	8	6	6	7	7
Cynodon dactylon II	Marin, N.L., Mexico (1994)	EDCP	67	72	67	71	69
		a, %	45	55	47	52	50
		b, %	35	28	34	31	32
		c, % h^{-1}	7	5	6	6	6
Dichanthium annulatum	Marin, N.L., Mexico (1994)	EDCP	46	44	45	65	50
		a, %	30	22	20	52	31
		b, %	27	41	43	22	33
		c, % h^{-1}	6	5	6	7	6

Dichanthium annulatum	Linares, N.L., Mexico (1998-99)	EDCP	56	71	67	73	67
		a, %	34	41	55	49	45
		b, %	33	42	17	33	31
		c, % h^{-1}	5	5	5	5	5
Panicum coloratum	Linares, N.L., Mexico (1998-99)	EDCP	55	60	56	67	60
		a, %	24	38	49	48	40
		b, %	40	47	10	29	31
		c, % h^{-1}	3	3	4	5	4
Rhynchelytrum repens	Teran, N.L., Mexico (2001-02)	EDCP	64	53	55	59	58
		a, %	53	41	44	48	47
		b, %	14	17	14	15	15
		c, % h^{-1}	4	5	4	5	5

EDCP = calculated with a ruminal rate of passage of 2 % h^{-1};
a = is an intercept representing the portion of the crude protein solubilized at the beginning of digestion (t = 0):
b = is the portion of crude protein slowly degraded in the rumen;
c = is the constant rate of disappearance of the fraction b; Obtained from:
Raiírez et al. (2002); Ramirez et al. (2003); Ramirez (2003); Ramirez et al. (2004); Ramirez et al. (2005);

Table 6.11. Effective degradability of crude protein (EDCP; % dry matter) and the characteristics of *the in situ* digestibility of crude protein of the hybrid buffelgrass Nueces and five new genotypes of buffelgrass collected in different dates in Teran county of the state of Nuevo Leon, México

Genotypes	Concept	Dates of collection			
		Aug.1999	Nov. 1999	Nov. 2000	Jun. 2000 (Fertilized)
Cenchrus ciliaris Nueces	EDCP	66	64	71	46
	a, %	41	47	50	35
	b, %	24	24	37	23
	c, % h^{-1}	4	4	3	5
Cenchrus ciliaris 307622	EDCP	64	63	67	45
	a, %	46	40	47	30
	b, %	28	41	37	25
	c, % h^{-1}	4	4	4	5
Cenchrus ciliaris 409252	EDCP	59	58	55	44
	a, %	39	37	30	40
	b, %	31	30	44	18
	c, % h^{-1}	4	4	4	5
Cenchrus ciliaris 409375	EDCP	64	53	59	45
	a, %	47	32	35	36
	b, %	33	30	40	21
	c, % h^{-1}	3	4	4	5

			Winter	Spring	Summer	Fall
Cenchrus ciliaris 409460	EDCP		55	64	67	44
	a, %		37	42	46	32
	b, %		35	42	34	20
	c, % h^{-1}		4	3	4	5
Cenchrus ciliaris 443	EDCP		65	70	64	59
	a, %		49	54	43	41
	b, %		26	25	34	30
	c, % h^{-1}		4	4	4	4

EDCP = calculated with a ruminal rate of passage of 2 % h-1
a = is an intercept representing the portion of the crude protein solubilized at the beginning of digestion (t = 0):
b = is the portion of crude protein slowly degraded in the rumen; c = is the constant rate of disappearance of the fraction b;
Obtained from: García-Dessommes *et al.* (2003ab).

Without exception, the EDCP in leaves of cultivated grasses listed in Table 6.10 resulted 10% greater than the stems of the grasses. Moreover, during the winter (dry season) the difference between leaves and stems was more pronounced. Leaves of Common buffelgrass harvested in Linares county of the state of Nuevo Leon, Mexico had the highest EDCP; however. *Panicum coloratum* resulted with the lowest values.

Table 6.12. Effective degradability of crude protein (EDCP; % dry matter) and the characteristics of the in situ digestibility of crude protein of leaves and stems of stems and leaves of cultivates grasses collected in different counties and dates in the state of Nuevo Leon, Mexico

Grasses	Place and date of collection	Parts	Concept	Seasons				Annual
				Winter	Spring	Summer	Fall	mean
Cenchrus ciliaris Común	Linares, N.L., México (1998-99)	Leaves	DEPC	64	77	72	73	72
			a, %	39	51	62	47	50
			b, %	41	40	14	35	33
			c, % h^{-1}	4	5	5	4	4
		Stems	DEPC	38	74	63	85	65
			a, %	28	54	51	73	52
			b, %	20	29	17	16	20
			c, % h^{-1}	5	7	7	7	7
Cenchrus ciliaris Común	Marín, N.L., México (1998-99)	Leaves	DEPC	61	70	80	73	71
			a, %	37	54	57	52	50
			b, %	35	30	34	31	33
			c, % h^{-1}	3	3	3	3	3
		Stems	DEPC	44	66	68	64	61
			a, %	26	49	43	47	41
			b, %	27	37	37	26	32

			c, % h⁻¹	4	4	4	4	4
Cynodon dactylon	Linares, N.L., México (1998-99)	Leaves	DEPC	64	72	65	60	65
			a, %	39	51	50	44	46
			b, %	43	35	22	31	33
			c, % h⁻¹	4	4	5	3	4
		Stems	DEPC	44	75	72	67	64
			a, %	28	60	60	54	50
			b, %	33	24	17	28	25
			c, % h⁻¹	3	4	4	3	3
Dichanthium annulatum	Linares, N.L., México (1998-99)	Leaves	DEPC	64	74	71	75	71
			a, %	45	38	57	46	47
			b, %	28	59	19	40	36
			c, % h⁻¹	6	4	7	7	6
		Stems	DEPC	43	72	57	60	58
			a, %	33	54	49	53	47
			b, %	13	23	10	9	14
			c, % h⁻¹	8	11	8	9	9
Panicum coloratum	Linares, N.L., México (1998-99)	Leaves	DEPC	58	67	59	62	61
			a, %	38	43	50	42	43
			b, %	38	47	13	33	33
			c, % h⁻¹	3	3	8	3	4
		Stems	DEPC	57	63	53	58	58
			a, %	43	44	46	46	45
			b, %	18	23	9	16	17
			c, % h⁻¹	12	12	14	10	12

EDCP = calculated with a ruminal rate of passage of 2 % h-1
a = is an intercept representing the portion of the crude protein solubilized at the beginning of digestion (t = 0):
b = is the portion of crude protein slowly degraded in the rumen; c = is the constant rate of disappearance of the fraction b.
Obtained from: Ramirez et al. (2001ab); Foroughbackhch et al. (2001); Ramirez et al. (2003); Ramirez (2003); Ramirez et al. (2005).

The EDCP of native grasses that grow in northeastern Mexico and southern Texas, USA, varied between species and seasons (Tables 6.11 and 6.12). *Panicum hallii* had the highest percentage and *A. longiseta* had the lowest annual mean value. Other grasses had intermediate values. In general, about 50% of the protein of the forage is degradable (100-EDCP) in the rumen. Therefore, this protein may represent the N that is digested and absorbed in the small intestine in the form of amino acids contributing to the N supply of the diet, that are used for the formation of protein in the host tissues.

Table 6.13. Seasonal tendency of the effective degradability of crude protein (EDCP; % dry matter) and the characteristics of the *in situ* digestibility of native grasses collected in Marin county of the state Nuevo Leon, Mexico 1994

Native grasses	Concept	Seasons				Annual
		Winter	Spring	Summer	Fall	mean
Aristida longiseta	EDCP	45	56	37	42	45
	a, %	25	29	32	43	32
	b, %	22	20	20	21	21
	c, % h^{-1}	6	5	6	8	6
Bouteloua gracilis	EDCP	54	57	54	60	56
	a, %	37	43	40	35	39
	b, %	28	25	24	39	29
	c, % h^{-1}	6	6	7	8	7
Cenchrus incertus	EDCP	50	54	49	65	54
	a, %	24	26	22	52	31
	b, %	44	45	47	20	39
	c, % h^{-1}	6	7	6	7	7
Hilaria belangeri	EDCP	40	53	61	65	55
	a, %	40	39	42	49	43
	b, %	21	33	33	25	28
	c, % h^{-1}	6	6	6	7	6
Panicum hallii	EDCP	54	76	60	59	62
	a, %	37	35	36	38	37
	b, %	30	26	25	37	30
	c, % h^{-1}	5	6	5	6	6
Setaria macrostachya	EDCP	49	63	60	69	60
	a, %	42	58	35	44	45
	b, %	29	18	23	33	26
	c, % h^{-1}	7	6	7	8	7

EDCP = calculated with a ruminal rate of passage of 2 % h-1
a = is an intercept representing the portion of the crude protein solubilized at the beginning of digestion (t = 0):
b = is the portion of crude protein slowly degraded in the rumen; c = is the constant rate of disappearance of the fraction b.
Obtained from: Ramírez *et al.* (2004)

Table 6.14. Seasonal tendency of the effective degradability of crude protein (EDCP; % dry matter) and the characteristics of the *in situ* digestibility of native grasses collected in Teran county of the state of Nuevo Leon, Mexico in 2001 and 2002

Native grasses	Concept	Seasons				Annual mean
		Winter	Spring	Summer	Fall	
Bouteloua curtipendula	EDCP	61	50	55	61	57
	a, %	49	36	41	48	44
	b, %	16	18	19	15	17
	c, % h^{-1}	5	5	4	5	5
Bouteloua trifida	EDCP	56	54	66	72	62
	a, %	44	43	58	63	52
	b, %	15	15	11	11	13
	c, % h^{-1}	5	5	5	4	5
Brachiaria fasciculata	EDCP	69	62	86	69	72
	a, %	58	52	79	56	61
	b, %	14	13	10	14	13
	c, % h^{-1}	5	5	4	4	5
Digitaria insulares	EDCP	67	48	81	75	68
	a, %	58	35	75	69	59
	b, %	13	16	9	8	12
	c, % h^{-1}	4	4	4	4	4
Chloris ciliata	EDCP	65	48	77	75	66
	a, %	57	29	67	63	54
	b, %	12	26	18	15	18
	c, % h^{-1}	4	5	4	4	4
Leptochloa filiformis	EDCP	69	62	62	72	66
	a, %	58	46	46	58	52
	b, %	14	23	23	19	20
	c, % h^{-1}	4	4	4	4	4
Panicum hallii	EDCP	57	57	70	74	65
	a, %	45	45	57	54	50
	b, %	17	17	19	13	17
	c, % h^{-1}	4	4	3	3	4
Panicum obtusum	EDCP	50	73	83	76	71
	a, %	39	50	72	64	56
	b, %	15	58	19	15	27
	c, % h^{-1}	4	4	3	4	4
Panicum unispicatum	EDCP	52	46	66	65	57
	a, %	38	34	52	51	44
	b, %	20	16	19	18	18
	c, % h^{-1}	4	4	4	4	4
Setaria grisebachii	EDCP	68	76	86	77	77
	a, %	58	68	74	68	67
	b, %	14	12	17	12	14

	c, % h⁻¹	4	4	3	4	4
Setaria macrostachya	EDCP	64	65	83	68	70
	a, %	55	53	73	59	60
	b, %	11	13	15	13	13
	c, % h⁻¹	4	4	3	4	4
Tridens eragrostoides	EDCP	68	74	74	68	71
	a, %	59	64	64	59	62
	b, %	12	15	15	13	14
	c, % h⁻¹	4	4	4	4	4
Tridens muticus	EDCP	58	51	73	61	61
	a, %	47	39	59	51	49
	b, %	15	15	20	14	16
	c, % h⁻¹	4	4	3	4	4

EDCP = calculated with a ruminal rate of passage of 2 % h-1

a = is an intercept representing the portion of the crude protein solubilized at the beginning of digestion (t = 0):

b = is the portion of crude protein slowly degraded in the rumen; c = is the constant rate of disappearance of the fraction b.

Obtained from: Cobio-Nagao (2004).

The native grass from Teran county of the state of Nuevo Leon, Mexico *Setaria grisebachii* had the highest EDCP percentage, but *B. curtipendula* and *P. unispicatum* were lowest (Table 6.12). In addition, the EDCP was variable between grasses and between seasons. The differences in EDCP mostly depend on the state of maturity and crude protein content of grasses. Table 6.12 shows that during wet seasons (summer and fall), grasses the highest EDCP value; conversely, during the dry season (winter and spring) grasses had the lowest values.

CHAPTER 7

Intake of grasses

Introduction

The most important variables that affect animal intake are: animal factors (physiological state, productive potential), food features (quality, quantity and plant species) and environmental aspects (Temperature and rainfall). The knowledge of these elements helps to better understand the interaction between plant, animal and environment. Food consumption is one of the best-regulated homeostatic mechanisms of the animal organism. Intake regulation occurs at different levels. For example, the animal has to balance the acquisition of nutrients to meet the daily and seasonal metabolic demands regulating the consumption between meals; in addition, has to observe various environmental conditions (photoperiod and food availability, filled rumen and nutrient uptake, body fat and body energy and nutrient requirements). These experiential and control mechanisms have to be incorporated hierarchical allowing maintenance of energy balance in different nutritional environments.

Factors in grasses that influence ruminant intake

Many factors affect grass intake being the smell usually the most important. Ruminants may reject feed without tasting it. For instance, the smell of manure reduces the intake of grass by cattle, but if the grass around the feces is cut and transported, the animal will consume it quickly. This behavior has possibly developed to protect the ruminant against infestation by intestinal parasites. Unpalatable pasture appears to emit volatile materials, since animals have rejected it without tasting. Moreover, ruminants will only consume small amounts of moldy feeds and unhealthy plants, such as rust-affected grasses, even under pen feeding. Furthermore, grasses that are dirty tend to produce irritation of the nose and eyes of ruminants and reduce feed intake. Inhibiting these materials increases intake and has important application in developing countries. Some producers add water to chopped wheat straw prior to feeding, which appears to increase intake of straw. In addition, chopping straw into short lengths produce increase intake of the straw. Fine grinding and pelleting increases intake of straw but has little applicability in developing countries because of the high energy costs associated with this form of processing.

Many reasons have been proposed as regulators of voluntary feed intake. In some circumstances, the inherent assumption has been that a factor performed independently and exclusively of other mechanisms. A knowledge of ruminant digestive anatomy aids in understanding both the ecological niche and the feeding behavior of the ruminant animal. Factors controlling ruminant intake should be assumed to function with multiple interactions. A number of feedback regulators such as distension, protein, and energy should be considered in the context of their interacting regulatory effects when attempting to predict intake. Behavioral aspects also influence voluntary feed intake through associations formed via post-ingestive feedback. Ruminants can learn to identify particular feeds and alter intake based on experiences. An integrated approach is proposed as a means of understanding ruminant feed intake regulation and eventually to improving prediction of intake. Empirical mathematical methods are likely to be fundamental to developing understanding and models of feed intake because of the difficulty of studying the central nervous system. Even these problems, knowledge of theoretical feedbacks has already been used to develop practical ruminant feeding strategies.

The ruminant stops eating because of physical or metabolic limitations, thus, the animal has to decide to what extent the disadvantages or deficiencies or excesses of certain nutrients overweight more than the benefits of trying to satisfy the energy requirements of the animal, of which are believed to be the moving forces of intake.

Physiological affecting intake regulation

The control of intake is multi-factorial. It depends, at the same time, on grasses characteristics in relation to the gut capacity, to the ruminant requirements and nutrient concentrations of grasses, to the post-ingestive feedback of the intake and the learning process, to the morphological characteristics of grazed plants, and on the environment such as climate, abundance and frequency of feed resources, etc.

Role of the ruminal fill

The fill gastrointestinal track capacity, in relation to grasses characteristics, may be considered as a key factor of regulation of voluntary feed intake. Intake appears limited by the maximal volume that the digestive tract can reach, even if herbivores are able to gradually adjust the volume of their rumen and to increase the transport rate of digesta when the quality of grasses decreased. This has been confirmed by the introduction, into the rumen water filled bags and artificial fibers. The more the ruminal content is bulky, in volume or in weight, the more the intake decreases with or without digestibility modification. It has been confirm that ruminal fill can affect grazing behavior in terms of bite mass, bite depth and bite area. In this manner, the ruminal fill affects the short-term intake.

In relation to the ruminal fill, grass dry mater content can influence the voluntary intake. If dry matter of grasses is lower than 20%, as in young grazed grass, the volume of water in the rumen increases and has depressive effect on the intake level, this in spite of a high forage digestibility. The age of plant is also a factor of variation. When plant mature, protein content decreases, cell walls and tissues lignification increase with, as a consequence, an increase of forage retention time in the rumen, limiting voluntary intake. The daily forage intake of lactating dairy cows decreases by 8.4% when comparing short and high age of grass regrowth. This was confirmed that the level of dry matter intake was negatively correlated to the hemicellulose and cellulose (NDF) content but as described by the low coefficient of correlation ($r = -0.65$ and $= -0.31$), NDF alone appears as a bad predictor of intake.

Main factors related in diminishing feed intake

There are three basic stimuluses associated with digestion and metabolism that rise from food pursuing and ingestion, and which either only or in combination inhibit the feeding centers of the hypothalamus and therefore limit feed intake. These are:

1. Absorption and metabolism of nutrients: when the products of digestion are imbalanced to meet a particular productive function, there will be an excess of C2-energy, which must be expended as heat. The animal reduces its feed intake as a consequence of this imbalance, particularly in warm climates
2. Distension of the gut: although distension of the tract can limit intake, this has often been overemphasized and with many feeds deficiencies of nutrients (mainly amino acids) are the first limiting factors
3. Fatigue: ruminants become fatigued in seeking, ingesting, chewing and ruminating their ingesta. In animals given fibrous feeds, twice as much feed passes through the rumination cycle as is eaten, which implies that rumination may be a major cause of fatigue. Feed intake is restricted by the need to ruminate and the time taken for this.

In ruminants on high-energy diets, the rate at which VFA are absorbed from the rumen may limit feed intake. This was demonstrated in sheep by infusing VFA into the rumen at any time, sheep began to eat. Infusion of VFA always reduced the meal size. Infusing butyrate had much less effect on the size of the meal than infusing either propionate or acetate.

Grazing behavior of ruminants

Grazed forage is the cheapest feed source available for cattle and sheep. However, grazing animals often fail to succeed their production potential because voluntary feed intake is usually lower than what may be achieved when offering conserved and processed feeds. Under grazing conditions, daily feed intake is the product of intake rate and grazing time, where intake rate may be considered as the product of intake per bite (bite mass) and bite rate, and grazing time is the product of mean meal duration and number of meals.

Grass condition, bite mass and bite rate

Under temperate pasture conditions, bite mass is very highly correlated with leaf area index (LAI, leaf area per unit ground area) or green leaf mass. However, although the precise relationship may be modified where grasses differ in their leaf to stem ratios or population densities, grass surface height (GSH) is a useful, practical method of applying the principles of forage growth and utilization originally based on LAI. Thus, it was found that as GSH is increased, bite mass increases, which in turn has a profound influence on bite rate and, as a consequence, intake rate.

Effects of physiological condition of the ruminant

The intake rate and grazing time may be varied by grazing animals to regulate daily intake, in response to their nutrient requirements. E.g., lactating sheep and cattle might be revealed to increase their intake rates by 10% and 19%, respectively, compared with non-lactating individuals grazing the same grasses. However, the main approach of sheep or cattle for meeting the increased nutrient demands of lactation is to increase total daily grazing time. Therefore, lactating cows and sheep have been demonstrated to increase the time spent grazing each day by 22% and 29%, respectively, compared with non-lactating counterparts. Because of increased intake rate and grazing time, daily intake by sheep and cattle might be improved by up to 40% in response to lactation.

Handling time

It has been suggested that the higher intake rate by cattle, compared to sheep, may not be due solely to the greater dental arcade size and hence bite mass; however, that the main factor influencing intake rate is the time required to take a bite plus the time taken to masticate the grass in that bite. Moreover, rumination may also be considered as a handling cost, which although it will not directly affect intake rate, might affect total intake by reducing the time available for grazing. It appears that, lactating cows require less time per unit of forage ingested than heifers, which consecutively, require less time than lactating ewes. Moreover, while sheep take less time per unit of clover ingested than grass, heifers take similar times. Both ewes and heifers have shorter rumination times per unit of clover ingested than of grass.

Nutrient intake by sheep grazing on a buffelgrass pasture

A year-round grazing system proves to be the most efficient management practice for the active conservation of valuable natural dry grasslands. Sheep, because of their body size and low nutritional requirements, are highly valued in conservation grazing programs. However, the nutritional value of a dry grassland varies greatly, which produces a number of adverse effects in sheep, especially regarding appropriate energy-protein balance in animal diet. Therefore, the nutritional well-being of managed sheep is likely to be disturbed, which requires supplementary feeding of animals. Buffelgrass pastures in northeastern Mexico also are affected by climate conditions, and nutrient content is reduced during the winter and drought.

In a study carried out on a buffelgrass pasture located in northeastern Mexico, the intake was estimated by the total fecal collection technique. Montly, five mature castrated lambs (Pelibuey X Rambouillet), weighing about 39 kg of approximately one year of age, were fitted with fecal collection bags, and grazed with the esophageal fistulae lambs. Fecal output was sampled by weighing feces and intake was calculated as:

Fecal output (g per day)/(1 - IVOMD).

The OM, crude protein (CP), neutral detergent fiber (NDF), acid detergent fiber (ADF), macro and trace mineral element intakes of lambs were estimated (Table 7.1).

Table 7.1. Live weight and nutrients intake by range sheep on a buffelgrass (*Cenchrus ciliaris*) pasture

Concept	Jan.	Feb.	Mar.	Apr.	May.	Jun.	Jul.	Aug.	Sep.	Oct.	Nov.	Dec.
Live weight, kg	39.2	34.1	36.7	39.1	37.6	37.2	39.6	36.7	39.0	41.0	43.1	42.0
OMI												
kg day^{-1}	0.7	0.8	0.6	0.9	1.1	1.0	1.1	0.7	0.7	0.7	0.7	0.9
g kg$^{0.75}$ day^{-1}	47.8	60.0	43.7	58.5	72.9	68.2	69.3	46.4	44.0	41.0	39.8	53.5
DEI[a]				—								
Mcal day^{-1}	1.2	0.9	1.0	1.5	2.7	2.0	2.1	1.0	1.2	1.2	1.5	1.5
Cal.kg$^{0.75}$ day^{-1}	75.7	59.5	65.8	98.1	183.2	136.6	138.0	65.3	79.3	73.4	87.2	91.3
CPI[a]												
g day^{-1}	90.1	108.7	94.0	138.6	145.2	156.1	189.7	118.7	93.1	90.4	88.1	100.0
g kg$^{0.75}$ day^{-1}	5.8	7.7	6.7	9.3	9.8	10.5	12.3	6.9	6.6	5.6	5.2	6.1
NDFI[a]												
g day^{-1}	590.4	644.1	511.7	699.4	819.3	817.2	974.3	561.9	617.2	593.7	757.0	753.3
g kg$^{0.75}$ day^{-1}	38.2	41.6	36.9	46.9	55.7	52.2	63.2	38.5	43.9	36.6	44.5	46.4
ADFI[a]												
g day^{-1}	418.3	402.4	363.2	431.5	576.3	392.9	446.0	361.6	362.2	362.9	412.9	395.4
g kg$^{0.75}$ day^{-1}	27.0	28.4	26.2	28.9	39.1	26.5	28.9	24.7	25.8	22.4	24.5	24.3

[a]Organic matter; OMI = organic matter intake; DEI; digestible energy intake; CPI = crude protein intake; NDFI = neutral detergent fiber intake; ADFI = acid detergent fiber intake.

The organic matter intake (OMI) of lambs was different among months (Table 7.1); Lambs consumed an annual mean of 816.6 g day^{-1}. The highest months were May, June and July (late spring and early summer). Lowest months were January and March (winter season). The OMI per kg$^{0.75}$ body weight (BW) was also different among sampling periods. Annual mean was 54.0 g kg$^{0.75}$ day^{-1}. Those months where OMI was highest coincided with the highest consumption of edible shrubs (annual mean = 14%) and forbs (annual mean = 1.0%) by sheep. In addition, OMI was negatively (r = - 0.64; P < 0.05) affected by grass consumption, and positively (r = 0.66; P < 0.05) by shrub consumption.

Daily digestible energy intake (DEI) of lambs was not uniform during the year (annual mean of 1469.3 kcal, or 96.1 kcal DE kg$^{0.75}$). The highest DEI occurred during May and the lowest during February. Growing lambs grazing in the same pasture required 105 kcal DE per $kg^{0.75}$ per day for maintenance, and a lamb weighing 25 kg and gaining 100 g day^{-1} required 1829.3 kcal DE $kg^{0.75}$ day^{-1}. It appears that lambs consumed sufficient DE to meet their maintenance requirements and gain weight only during May, June and July.

The crude protein intake (CPI) varied among months (annual mean 117.7 g day^{-1}). During July, lambs consumed the highest amount of CP. However, the lowest amounts were consumed in November. The same pattern was observed for CPI $kg^{0.75}$. It seems that lambs consumed forage with CP in amounts to satisfy their maintenance requirements (47.2 g kg^{-1}). It was reported that a 25 kg growing lamb gaining 100 g per day requires about 184 g CP day^{-1}. Therefore, only in July did lambs consume forage containing enough CP to gain weight.

Highest levels of neutral detergent fiber intake (NDFI) occurred during May, June and July. It seems that the type of plant composition on diets of sheep affected NDFI. High consumption of shrubs by lambs increased (r = 0.60; P < 0.05) NDFI, while high consumption of grasses was reduced (r= - 0.58; P< 0.05) NDFI (Table 7.1). The same pattern was observed with acid detergent fiber intake (ADFI). In general, lambs increased their fiber consumption in those months when the amount of shrubs in their diets increased. Thus, grassland rich in various plant species is more stable and more valuable from the ecological and probably also from the animal nutrition viewpoints.

The botanical composition of monthly diets of sheep affected nutrient intake. In those months (May, June and July) when the level of shrubs increased in the diets, OMI, CPI, NDFI and ADFI increased. In the remaining months, when lambs increased the level of grasses in their diets, the reduced nutrient intake became a problem in meeting the nutrient requirements for grazing sheep. Native shrubs growing in buffelgrass pastures are reported to be important diet components for grazing ruminants, especially during the fall and winter seasons.

Minerals intake by sheep grazing on a buffelgrass pasture

Ruminants that depend primarily on forage to satisfy mineral requirements have suffered deficiencies in one or more minerals because concentrations of mineral elements in forage depend upon interactions of factors such as soil, plant species, stage of maturity, yield, pasture management and climate. Moreover, increased lignification and cell wall content may reduce the total minerals consumed. In a grass nutrition study, the mineral intake was assessed, in range sheep consuming buffelgrass in a pasture located at northeastern Mexico.

The amount of K (Table 7.2) consumed by sheep was different among sampling periods (annual mean = 19.5 g per day). During the summer months (July, August and September, lambs had the highest K consumption; however, the lowest intake occurred during winter months. The K requirement of sheep appears to be about 0.5% of the diet DM day^{-1}). In this study, lambs consumed sufficient amounts of K, in every month, to meet and exceed their requirements. Moreover, K concentration in forage collected from the study area averaged 20.1% of the DM.

The Ca intake of lambs was dissimilar among months (annual mean 4.8 g day^{-1}). During May Ca intake was highest; the lowest Ca consumption occurred in January and February (Table 7.2). It seems that sheep require 0.20-0.80% of Ca in diet. In this study, lambs consumed amounts of Ca to meet their requirements during all months except January and February, when lambs selected forage with marginal deficiencies of Ca to meet their requirements.

The amount of Mg consumed by lambs was not uniform throughout the year (Table 7.2). Annual mean value was 2.I g day^{-1}. During May, lambs had the highest intake and during February was lowest. The exact requirement of Mg for sheep is unknown, although forage containing 1.5 g kg^{-1} of Mg in diet DM is considered adequate for adult range ewes. In this study, lambs consumed only during April, May, June, July and December amounts of Mg to meet their requirements; other months were marginal insufficient. It has been reported that Mg becomes a problem most frequently when ruminants are just entering lactation and placed in pastures containing less than 0.2 g Mg kg^{-1} in the diet DM.

Table 7.2. Monthly macrominerals intake by sheep on a buffelgrass (*Cenchrus ciliaris*) pasture

Element[a]	Jan.	Feb.	Mar.	Apr.	May.	Jun.	Jul.	Aug.	Sep.	Oct.	Nov.	Dec.	Mean
K, g.day^{-1}	13.6	14.2	14.5	14.5	14.5	21.4	34.7	18.2	33.0	27.6	12.9	13.6	19.5
Ca, g.day^{-1}	1.7	1.9	3.9	3.6	8.9	5.8	5.9	5.0	5.2	3.9	6.8	5.5	4.8
Mg, g.day^{-1}	0.6	0.5	3.1	2.7	5.3	2.1	2.1	1.2	1.2	1.0	1.0	1.6	2.1
Na, g.day^{-1}	5.4	5.6	7.1	4.5	5.1	6.4	6.2	4.6	12.5	8.3	7.3	5.0	6.5

[a]Dry matter.

Apparently, forage consumed by lambs in some months did not contain sufficient Cu to meet their requirements (5-10 mg kg^{-1}), especially during the fall and winter seasons (Table 8.3). Among the trace elements evaluated in the forage collected in the buffelgrass pasture, Cu was the most critical element to meet sheep requirements. Only during August and September did levels of Cu exceed requirements for grazing ruminants. A wide number of symptoms are associated to Cu deficiencies among cattle, and their diversity may be linked to complex interactions involving other minerals. Some of the clinical signs include: bleaching of hair, nervous symptoms (ataxia) in calves whose dams experienced deficiency during pregnancy, lameness, and swelling of joints. Therefore, range managers have to monitoring the Cu status of animals and be more attentive to management of their mineral programs.

The Fe consumed by lambs was different among months. In the summer, lambs had the highest consumption of Fe, compared to other seasons (Table 7.3). Throughout the year, sheep consumed diets containing sufficient amount of Fe to meet growing ruminants (50 mg kg^{-1} in the diet DM). Similar response was reported in a study, in which mineral content in seven range grasses were assessed; it was found that the mean Fe content of the grasses was 194.4m g kg^{-1}, and each of the grasses exceeded or equaled beef cattle forage requirements for all months sampled. High levels of Fe could potentially lower Cu availability and exasperate management problems associated with Cu deficiencies.

The Mn intake by lambs, in this study, was not uniform among sampling periods (Table 7.3). It has been established that sheep require 20-40 mg Mn kg^{-1} of diet DM. Only during February, did lambs not consume amounts of Mn to meet their requirements. Similar response has been found in seven native grasses collected monthly in US. Only cheatgrass (*Bromus tectorum* L.) contained adequate concentrations of Mn early in the growing season, and it exceeded beef cattle requirements for all but the last months of both years. Other grasses, however, like bluebunch wheatgrass (*Agropirum spicatumor*) bottlebrush squirreltail (*Sitanion hystrix*), consistently supported adequate levels of Mn late in the year, so cattle consuming a diverse diet could probably ingest sufficient Mn on a season long basis.

Grazing sheep require 20-30 mg Zn kg^{-1} of diet DM. In this study, amounts of Zn consumed by lambs met, and in some months exceeded, their requirements (Table 7.3). In most months, forage collected in the buffelgrass pasture contained amounts of Zn to meet sheep requirements. It appears that Zn content declines as buffelgrass mature progressed (January through April). Similar tendency was reported in other study, in which evaluated the Zn content of the grasses *Poa sandbergii, Bromus tectorum, Sitanion hystrix, Agropyron spicatum, Festuca idahoensis, Stipa thurberiana,* and *Elymus cinereus,* exhibited a nearly linear or curvilinear decline of Zn as the seasons progressed. Moreover, none of the grasses sampled met the required concentration of Zn for beef cattle forage (30 mg kg^{-1}) for any sampling period in either year. Zinc deficiencies can cause parakeratosis (inflamed skin around nose and mouth), stiffness of joints, alopecia, breaks in skin around the hoof, and retarded growth.

Table 7.3. Monthly microminerals intake by sheep on a buffelgrass (*Cenchrus ciliaris*) pasture

Element[a]	Jan.	Feb.	Mar.	Apr.	May.	Jun.	Jul.	Aug.	Sep.	Oct.	Nov.	Dec.	Mean
Cu, mg day^{-1}	2.5	0.7	5.3	9.0	8.3	10.6	12.5	7.8	7.0	3.6	4.1	4.0	6.3
Fe, mg day^{-1}	307.6	198.4	306.2	286.4	349.6	365.5	372.5	252.4	299.6	240.8	319.3	344.4	303.4
Mn, mg day^{-1}	39.8	15.5	66.8	51.6	59.3	114.6	118.9	41.9	71.8	56.0	52.0	76.9	63.8
Zn, mg day^{-1}	27.5	25.6	22.6	27.5	30.3	71.7	381.8	220.9	250.9	204.0	223.3	75.2	130.0

[a]Dry matter.

Moreover, shrubs, in this study, during the spring and summer, provided adequate levels for sheep production of macro and trace elements in the forage consumed for lambs. The exception was Cu, which in most months of the year was marginally deficient for sheep requirements. Buffelgrass pastures in northeastern Mexico that grow mixed with native range shrubs may provide a better nutrition for grazing ruminants than pastures with only buffelgrass, especially during spring and summer seasons, and maintain body condition during fall and winter seasons.

Grazing cattle select from a variety of available forages. Early in the growing season, cattle may select up to 80% of their diet from a single grass, but their diets become more diverse as forages mature. Therefore, a great portion of their intake is derived from other sources, and there may be forbs and shrubs available that can also help rectify dietary deficiencies. In addition, livestock have many mechanisms for either conserving, recycling, mobilizing or buffering mineral or nutrient balances within their systems, and these mechanisms allow them to endure short-term deficiencies without harsh effects. However, lactating animals may not benefit themselves of many of these mechanisms when grazing deficient forages.

In intensively managed pasture, many mineral deficiencies can be rectified by treating the land with a required element or altering pH of the soil to enhance mineral availability for growing forages. Other options include either oral treatments or injections for stock. With largescale feedlot situations, frequent ration sampling and custom supplement formulations may be mixed on even a daily basis to accommodate the dynamics of variable quality in feed supplies. In most extensive rangeland systems, however, these solutions are economically and/or logistically impossible, and the only recourse is to supply free-access supplements in either block or loose form.

Prediction of intake of grasses

Intake is particularly important because it is a primary factor controlling ruminant production from forages. The intake of a large number of grasses at different growth stages is difficult for most research laboratories due to of the huge area of land required to produce the amount of forage required for that activity. Thus, it may be necessary to predict the intake from the laboratory analyses of small samples collected in the field. Intake has been shown to be most closely related to neutral detergent fiber (NDF) concentration. However, the relationship between NDF and intake varies depending on dominant environmental conditions such as soil fertility, soil moisture, temperature, level of irradiance, wind, relative humidity and evaporative demand and must be determined for each study area. For example, soil salinity has been reported to increase organic matter digestibility (OMD) and decrease NDF and may influence digestibility and intake because of its effect on N and oxalic acid accumulation.

The development of a predictive grass dry matter intake (PDMI) model is important at both farm and research level. Because accurate estimation of PDMI enables a greater degree of accuracy in dairy cow nutrition. In addition, analysis of the relationships between pasture variables and animal performance conducted during model development characterizes the pasture characteristics affecting productivity. Moreover, the development of a PDMI prediction model will provide a means for the efficient evaluation of meat production performance potential of different grass cultivars, grassland management techniques, supplementary feeding strategies, and others. Furthermore, the model may be capable of assessing the implications for cattle grazing systems of future component research findings. And finally, dry matter intake is a key driver of methane emissions by ruminants or an accurate assessment.

Using the following predictive equation, the dry matter intake (DMI) was determined in different cultivated and native grasses cutting in different places of northeastern Mexico. The relationship between forage NDF and DMI:

$$\text{DMI (g kg BW}^{-1}\text{ day}^{-1}) = 86.5 - 0.09 \text{ NDF } (r2 = 0.87; \text{ RSD} = 1.16; P < 0.001)].$$

Predicted dry matter intake of cultivated grasses

It appears that, the PDMI of grasses was lower during summer and higher in winter. During spring and autumn, the PDMI remained the same. *Cenchrus ciliaris* collected in Linares county had the highest value and *Dichanthium annulatum* and *Panicum coloratum* were lowest (Table 7.4). Wet and warm days during winter season are common in these regions. This fact might be occurred that promoted that these grasses grow. The two varieties of *Cynodon dactylon*, in winter, resulted with the highest DMI than other grasses. Beef cattle consuming the grasses during summer time may have difficulties to support their productivity because their low DMI due to low quality of grasses. It appears that, a low quality feed is considered when cows will consume between 5.1 and 7.2 kg day^{-1} of grasses (1.8% and 2.0% of their body weight, on a dry basis, respectively). A quality feed is when cows will consume between 7.2 kg day^{-1} and 8.7 kg day^{-1} (2.0% and 2.2% of their body weight, on a dry basis, respectively). Moreover, a high quality feed is when cows will consume 10.5 kg day^{-1} (2.5% of their body weight, on a dry matter basis). Thus, outstanding PDMI was recorded in *Cenchrus ciliaris* in autumn collected in Teran county and the two varieties of *Cynodon dactylon* in winter collected in Marin county of the state of Nuevo Leon, Mexico (Table 7.4).

Table 7.4. Seasonal predicted dry matter intake (kg day^{-1}) of a 400-kg cow consuming cultivated grasses cut in different counties of the state of Nuevo Leon, Mexico

Grasses	Place and date of collection	Seasons				
		Winter	Spring	Summer	Autumn	Mean
Cenchrus ciliaris	Marin, N.L., Mexico (1994)	6.5	8.0	7.2	8.0	7.4
Cenchrus ciliaris	Teran, N.L., Mexico (2001-02)	5.4	8.3	3.6	10.1	6.9
Cenchrus ciliaris	Linares, N.L., Mexico (1998-99)	8.7	8.7	9.4	9.0	9.0
Cynodon dactylon	Linares, N.L., Mexico (1998-99)	5.4	8.7	4.7	6.5	6.3
Cynodon dactylon	Marin, N.L., Mexico (1994)	9.8	6.9	5.1	7.2	7.2
Cynodon dactylon II	Marin, N.L., Mexico (1994)	9.8	5.4	5.4	7.2	7.0
Dichanthium annulatum	Marin, N.L., Mexico (1994)	6.5	6.9	6.5	6.2	6.5
Dichanthium annulatum	Linares, N.L., Mexico (1998-99)	6.5	6.9	4.7	6.9	6.3
Panicum coloratum	Linares, N.L., Mexico (1998-99)	5.8	6.9	5.1	7.2	6.3
Rhynchelytrum repens	Teran, N.L., Mexico (2001-02)	8.3	9.8	8.0	8.3	8.6
Seasonal means		7.6	7.2	6.5	7.2	

The PDMI of unfertilized new genotypes of buffelgrass was higher in the cut of November of 2000 than other cutting dates with or without fertilization (Table 7.5). The 443 genotype had the highest predicted DMI during the two cuts (1999 and 2000) carried out during autumn and was lowest during summer (1999). Fertilization with 100 kg urea N ha^{-1} had not effect on predicted DMI of the genotype 443 collected in June 2000. There is evidence the lack of influence of fertilization on grasses in ruminants. Similar findings were observed in intake by steers on grass fertilized with either high or low levels of nitrogen. Moreover, it was observed no differences in intake by sheep of *Phleum pratense* fertilized at 45 or at 134 kg of N ha^{-1}. In addition, it was noted that fertilization of *Festuca arundinacea* had no effect on PDMI by sheep. Furthermore, it was reported that neither level nor source of N fertilizer had a significant effect on level of *Dactylis glomerata* intake by sheep.

Table 7.5. Predicted dry matter intake (kg day^{-1}) of a 400-kg cow consuming new genotypes of *Cenchrus ciliaris* cut in Linares county of the state of Nuevo Leon, Mexico

Genotypes	Cuts			
	Aug. 1999	Nov. 1999	Nov. 2000	Jun. 2000 (Fertilized)
Cenchrus ciliaris Nueces	8.0	9.0	8.7	8.7
Cenchrus ciliaris 307622	9.0	9.0	9.4	9.0
Cenchrus ciliaris 409252	8.3	9.8	9.4	9.4
Cenchrus ciliaris 409375	8.3	9.8	9.4	8.3
Cenchrus ciliaris 409460	8.7	9.8	9.4	8.0
Cenchrus ciliaris 443	9.0	10.8	10.8	9.8
Means	8.7	9.8	9.4	9.0

The PDMI of unfertilized of 84 new genotypes of *Cenchrus ciliaris* was higher in both genotypes 409258 and 409459 with 10.1 kg day-1 (2.5% of body weight of dry matter basis) and the genotype 414499 was lowest with 6.5 kg day-1 (1.6% of body weight of dry matter basis; Table 7.6). The mean PDMI of the 84 grasses (8.3 kg day-1) represent 2.1% of body weight of dry matter basis. Thus, most of the 84 genotypes are feeds of good quality when cows will consume these grasses.

Mean of PDMI intake of leaf of the four grasses listed in Table 7.7 was 69% higher than that of stem. It has been reported that higher intake of the leaf fraction is associated (r = 0.74; P< 0.01) with a shorter retention time of dry matter in the reticulo-rumen, which appeared to be caused by the large surface of the leaf fraction initially available to bacterial degradation. Moreover, it was concluded that the higher voluntary intake and shorter retention time in the rumen of the leaf than of the stem fraction of grasses was associated with an apparent higher rate of digestion of NDF in vivo.

Table 7.6. Predicted dry matter intake (PDMI; kg day^{-1}) of a 400-kg cow consuming 84 new genotypes of *Cenchrus ciliaris* cut in Teran county of the state of Nuevo Leon, Mexico

Genotypes	PDMI	Genotypes	PDMI	Genotypes	PDMI	Genotypes	PDMI
202513	8.3	409185	8.7	409258	10.5	409449	8.3
253261	8.7	409197	9.4	409263	9.4	409459	10.5
307622	9.8	409200	6.2	409264	8.3	409460	8.7
364428	9.8	409219	10.1	409266	8.0	409465	7.6
364439	7.2	409220	9.0	409270	8.3	409466	7.2
364445	8.3	409222	7.2	409278	9.0	409472	8.7
365654	8.7	409223	7.6	409280	9.4	409480	7.2
365702	8.7	409225	6.9	409300	7.6	409529	7.2
365704	9.0	409227	7.6	409342	9.4	409691	8.0
365713	7.6	409228	9.0	409359	8.3	409711	7.2
365728	7.2	409229	10.1	409363	10.1	414447	7.2
365731	6.9	409230	9.0	409369	9.4	414451	7.6
409142	7.6	409232	9.4	409373	9.8	414454	7.6
409151	8.7	409234	9.4	409375	8.7	414460	7.6
409154	9.4	409235	9.4	409377	9.0	414467	7.6
409155	9.0	409238	10.1	409381	9.0	414499	6.5
409157	9.0	409240	9.4	409391	7.6	414511	7.2
409162	9.8	409242	7.2	409400	7.6	414512	8.0
409164	8.0	409242	9.4	409410	8.0	414520	7.2
409165	9.8	409252	7.6	409424	9.4	414532	9.0
409168	9.4	409254	8.3	409448	7.2	443	8.7
Mean of all genotypes							8.3

Predicted dry matter intake of native grasses

The PDMI of the native grasses Brachiaria fasciculata and Panicum obtusum was highest and Tridens muticus was lowest (Table 7.7). Thus, cows on the former grasses will consume a high quality feed.

In general, cows on the rest of the grasses, in all seasons, will consume a quality diet. The PDMI of all native grasses during spring was highest and in winter was lowest. Because the performance of livestock grazing grasses is directly related to the quantity and quality of grass consumed; therefore, all native grasses listed in Table 8.7 might support a good productivity of the animals that consume them.

Table 7.7. Predicted dry matter intake (PDMI; kg day–1) of a 400-kg cow consuming leaves and stems of four cultivated grasses in Linares county of the state of Nuevo Leon, Mexico

Grasses	Places and dates of collection	Parts	Seasons				Annual
			Winter	Summer	Summer	Fall	mean
Cenchrus ciliaris	Linares, N.L., Mexico (1998-99)	Leaves	6.5	12.3	4.7	11.9	9.0
		Stems	2.9	7.6	2.9	8.0	5.4
Cynodon dactylon	Linares, N.L., Mexico (1998-99)	Leaves	6.9	5.8	5.1	7.2	6.2
		Stems	5.1	8.3	4.4	5.8	5.8
Dichanthium annulatum	Linares, N.L., Mexico (1998-99)	Leaves	7.2	8.3	5.8	8.0	7.2
		Stems	4.4	5.1	4.4	4.7	4.7
Panicum coloratum	Linares, N.L., Mexico (1998-99)	Leaves	5.8	8.7	6.5	10.1	7.6
		Stems	4.0	5.4	3.6	4.7	4.4

Rangeland grasses are known to yield various quantity of forage depending on the prevailing environmental conditions. The differences in biomass yields in grasses can also be attributed to their growth characteristics, morphological and physiological properties and competitive advantage of the individual grass species constituting the mixture in the grassland. In rangelands located in northeastern Mexico the seasonality of rainfall (784 mm year-1), and high temperatures in spring and summer and low in winter (annual mean = 22.4 °C; evaporation 1622 mm), are major negative influences on nutritional yield of native grasses.

In all seasons, PDMI of the grasses listed in Table 7.8 was low, thus, range sheep on all grasses will consume low quality feeds throughout the year, except for *Panicum hallii* that sheep on this grass will consume a quality feed in all seasons of the year. Native grasses in Marin, county are growing under more severe environmental conditions. Average precipitation, by month, ranges from 200 to 800 mm, but 6 to 8 months of the year are relatively dry. Transpiration and evaporation far exceed input from precipitation. The annual transpiration/precipitation (T/P) ratios exceed 23, so the area could be realistically classified as arid rather than semi-arid. High temperatures account for the elevated evaporative losses. Thus, the biomass yield of Marin is lower than in Teran county of the state of Nuevo Leon, Mexico

Table 7.8. Predicted dry matter intake (PDMI; kg day^{-1}) of a 400-kg cow consuming native grasses collected in Teran county of the state of Nuevo Leon, Mexico

Native grasses	Seasons and years of cut				Annual
	Winter 2002	Spring 2002	Summer 2002	Fall 2001	mean
Bouteloua curtipendula	8.7	8.7	8.3	6.2	8.0
Bouteloua trifida	7.6	9.4	7.2	7.2	8.0
Brachiaria fasciculata	8.7	11.9	12.3	13.0	11.2
Digitaria insularis	9.4	10.5	11.2	9.0	10.1
Chloris ciliata	9.4	9.4	8.0	7.6	8.7
Leptochloa filiformis	7.6	10.5	8.3	10.5	9.4
Panicum hallii	8.0	10.5	10.1	9.0	9.4
Panicum obtusum	8.0	11.2	10.8	10.8	11.2
Panicum unispicatum	9.4	11.6	9.8	10.5	10.1
Setaria grisebachii	8.3	9.0	12.6	8.3	9.8
Setaria macrostachya	8.7	10.1	11.9	8.3	9.8
Tridens eragrostoides	9.0	8.0	8.7	7.2	8.3
Tridens muticus	7.2	8.7	8.3	6.5	7.6
Seasonal means	8.3	10.1	9.8	8.7	9.4

Table 7.8. Predicted dry matter intake (PDMI; g day^{-1}) of a 40-kg sheep consuming native grasses collected in Marin county of the state of Nuevo Leon, Mexico

Native grasses	Seasons				Annual
	Winter	Spring	Summer	Fall	mean
Aristida longiseta	328	400	328	292	328
Bouteloua gracilis	220	544	508	688	508
Cenchrus incertus	580	796	580	760	688
Hilaria belangeri	472	760	508	760	616
Panicum hallii	724	976	832	1012	868
Setaria macrostachya	580	616	364	796	580
Means	472	688	508	724	580

Intake of sheep energy supplemented on a buffelgrass pasture

Successful sheep production on pasture depends on sufficient daily forage intake to meet sheep nutritional requirements; otherwise, deficiencies must be corrected through supplementation if animal performance is to be maximized. In a study, was determined the influence of level of energy supplementation on daily intake of lambs grazing in a buffelgrass pasture. During a period of 15 weeks, 40 lambs (Rambouillet X Pelibuey; 25 castrated males and 15 females) were randomly grouped among five energy treatment levels (five males and three females per treatment level). The energy supplement ranged from 0.8% to 2.0% of body weight (BW), adjusted weekly. After the morning supplement feeding, lambs were allowed to graze freely for 7-9 h in a buffelgrass (*Cenchrus ciliaris*) pasture. Forage intake was determined with the 25 male lambs (five lambs per treatment) which were harnessed the last 10 days of the study. Total fecal excretion was registered in the last 5 days. Organic matter intake (OMI) was calculated as:

OMI = OM fecal output (g per day) / (1 - IVOMD).

In vitro OM digestibility (IVOMD) was assessed from the fistula extrusa forage collected from six esophageally fistulated male lambs, which were grazing with harnessed lambs, in the same area and period. During 4 consecutive days, extrusa were collected, 2 days in the morning and 2 days in the afternoon. Fistula extrusa samples and feces were oven dried (60°C) and ground through a 1 mm screen in a Wiley mill before analysis. Samples then were analyzed for dry matter (DM), ash, crude protein (CP) and acid detergent fiber (ADF). Metabolizable energy (ME) content of samples was estimated from DE X 0.82. DE intake (DEI) and ME intake (MEI) were calculated by the following equation of:

DEI = (g day^{-1} of OMI) (kcal g extrusa OM) - (g day of fecal output) (kcal g feces)

The OMI (g kg$^{0.75}$) of male lambs increased with an increase in energy supplementation (Table 7.9). Total OMI increased as the energy level increased up to 2.0% of BW. Forage OMI of male lambs was not affected by energy supplementation. However, lambs receiving higher energy levels consumed lower amounts of forage. Similar responses were found in other studies who reported that individually penned Somali lambs fed Napier grass green chop increased DM intake from a low of 85.0 (0.4% BW level of energy supplementation) to a high of 96.6 g per kgo.75 (1.6% BW). The CPI of male lambs (Table 8.9) increased as energy supplement increased from 0.8% to 1.7% BW (9.8 to 14.4 g per kg0 75). No further increase was obtained from 1.7% to 2.0% BW. The ADFI and DEI were not affected by the level of energy supplementation.

Table 7.9. Intake (OMI) of energy supplemented lambs on a *Cenchrus ciliaris* pasture

Concept	Level of energy supplementation (% of live weight)				
	0.8	1.1	1.4	1.7	2.0
OMI g $kg^{0.75}$					
Supplement	19.4	26.3	30.7	34.7	38.6
Forage	53.1	53.8	52.8	51.8	52.5
Total	72.5	80.1	83.5	86.5	91.2
CPI, g $kg^{0.75}$	9.8	11.0	11.6	12.1	12.9
ADFI, g $kg^{0.75}$	23.8	24.4	24.2	21.9	24.3
DEI, kcal $kg^{0.75}$	170.9	171.1	199.7	197.5	210.8

OMI = organic matter intake.
CPI = crude protein intake.
ADFI = acid detergent fiber intake.
DEI = digestible energy intake.

CHAPTER 8

Macrominerals in grasses

Introduction

Macrominerals (Ca, Mg, N, Na, P, S and Si; Table 8.1) are required for virtually all the vital processes of the animal organism. Moreover, their requirement is greater than 1.0 mg kg^{-1}. A deficiency of essential macrominerals in animals results in abnormalities that can only be corrected supplementing the mineral deficiency. In addition to the requirements for functions of the mammals fed on pasture, ruminants are dependent on an adequate supply of minerals for optimal rumen microbial activity, and thus a better utilization of forage. The grasses provide an important source of minerals for ruminants. Under certain circumstances, they can provide adequate amounts of all essential minerals required by these animals; however, on other occasions, the grasses are deficient in one or more minerals and need to be supplemented for optimal animal performance and/or health. Severe mineral deficiencies occur in various degrees, but the marginal deficiencies are probably much more common. The marginal mineral imbalance can occur without clinical manifestation, but rather only a small decrease in metabolic functions, substantially affecting the growth, reproduction and health.

Table 8.1. Name, chemical symbol and properties of some macroelements

Name	Symbol	Atomic number	Atomic mass, g/mol	Densities to 20 °C, g/cm^3	Fusion point, °C	Ebullition point, °C
Calcium	Ca	20	40.078	1.54	839	1808
Potassium	K	19	30.0983	0.86	63.7	774

Magnesium	Mg	12	24.305	01.74	648.8	1107
Nitrogen	N	7	14.00664	1.17 g/l	-209.9	-195.8
Sodium	Na	11	22.989768	0.97	97.8	892
Phosphorus	P	15	30.973762	1.84	44 (P4)	280 (P4)
Sulfur	S	16	32.066	2.06	113	444.7
Silica	Si	14	20.0855	2.33	1410	2355

Factors affecting the mineral content of grasses

The concentrations of mineral elements in grasses depend on the interaction of various factors such as soil, forage species, maturity level, yield, grass management, and climate. In addition, body mineral composition can be affected the mineral composition of the grasses being consumed. Most of the deficiencies that occur naturally in herbivores are associated with specific regions and soil characteristics. The rate of absorption of minerals from the soil by grasses might be modified by the characteristics of having the soil drain and availability is reduced by increasing the soil pH. For most mineral elements, there are plants that serve as accumulator that contain extremely high levels of a specific mineral. As plants mature, their mineral content decreases because of a natural process of dilution and translocation of nutrients to the root system. In general, K, Mg, Na and Cl concentration decreases in direct proportion to the maturation of the plant.

Grass mineral availability

The ability to provide a grazing animal an adequate supply of minerals depends on the mineral content and its bioavailability. The bioavailability of a mineral is defined as the absorbed proportion of the ingested element, transported to its site of action, and physiological converted to its active form. Determining the concentration of most minerals in the forage is relatively simple; however, it is difficult to accurately measure their bioavailability to carry out a specific function in the animal. For this reason, the dietary concentration of a given mineral must be above the animal's requirement for a specific function if the mineral will be absorbed and utilized with maximum efficiency. The body always try to maintain mineral concentrations within a narrow range through homeostatic mechanisms of control that reduces absorption or increases excretion.

Biological requirements

The mineral requirements for adequate microbial fermentation in ruminants are usually lower than those required by the host (Figure 8.1). Microbes have no structural requirement of Ca or P, to that required by vertebrates for bone formation, but they need H, C, O, N, P and S that are the major components of the organic cellular composition. Proteins are the major source of N and S. Nitrogen and S ratio is about 12:1 for the composition of essential amino acids. The ruminal flora and consequently ruminants may require some elements in higher amounts than those required for other type animals. For example, Co is used as B12 by ruminal microbes, and Ni that are required by microorganisms is also a cofactor in ureases.

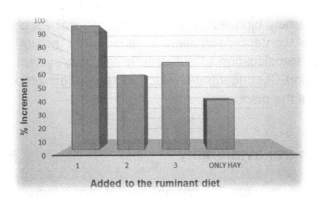

Figure 8.1. Effect of microbial population by the addition in the ruminant diet of 1) hay + minerals + vitamins + crude protein/nonprotein nitrogen, 2) hay + crude protein/nonprotein nitrogen, 3) hay + minerals + vitamins, and only hay.

Minerals in rumen fermentation

The P in ruminants is an element of relevant importance for proper metabolism and ruminal microflora health. Therefore, in ruminants, two types of P requirements have to be considered: one for the animal and other for rumen microorganisms. The P is part of the nucleic acids (DNA and RNA) is found in all bacterial cells. In bacterial cells in the rumen, the 10.3% of DNA and 9.64% of RNA are made up of this element. Most of the RNA from cells is found in the ribosome and ribosomal bacteria content is directly related to bacterial growth, and therefore, to the cellulolytic activity.

Zinc is essential for all living biological systems. The lack of availability of Zn in bacteria inhibits their multiplication, affecting the ability of cellulolytic bacteria to adhere to the cell wall of plant tissue and actively exercise their cellulolytic capacity. However, Mn is required for growth of most cells and exert an important role in the decarboxylation reactions of the tricarboxylic acid cycle. Furthermore, it has been shown that Mn stimulate CO_2 fixation in the production of succinic acid by rumen bacteria.

General functions of minerals in ruminant tissues

Minerals develop many functions that are directly or indirectly related to animal growth. They help in maintaining the rigidity of the bones and teeth, and represent an important part of proteins and lipids in the animal. In addition, minerals preserve the cellular integrity by osmotic pressures and are a component of many enzyme systems that catalyze metabolic reactions in biological systems. At least 15 elements are nutritionally essential for livestock. Main mineral nutrients are macrominerals such as Ca, P, Mg, K, Na, S, Cl, I, and trace elements such as Fe, Mo, Cu, Co, Mn, Zn, and Se.

Several factors affect the requirements of these minerals in supplements or food ingredients, including the interrelationships between minerals and other nutrients, consumption of mineral supplement, bred and adaptation of livestock.

The profiles of minerals in the soil and livestock tissues (blood, liver, bone and hair) help only to support the results obtained when deficiencies or poisoning in the results of analysis of fodder and water consumed by cattle are detected, which are the best indicators of deficiencies in grazing. In tissues, blood analysis provides a reliable retrospective in determining mineral deficiencies or excesses, but no more than those that provide the analysis of bone and liver, offering the advantage of its availability and ease of use without sacrificing the animal.

The animals have three primary sources for obtaining inorganic elements in livestock systems: 1) food, 2) water and 3) mineral supplements. Although plants can provide much of the necessary minerals, mineral supplementation is a necessary practice in good productive animals, depending on the type of animal production system. However, the mineral content of the feed depends primarily on the forage species, abundance of the element in the soil and persistent conditions for plant growth, and consequently on mineral uptake.

Macrominerals requirement by grazing ruminants

Many factors affect mineral requirements of grazing ruminants, including type and level of production, age and chemical form of the element in food, interaction with other minerals, mineral supplement consumption, bred and adaptation of the animal. Mineral requirements depend on the level of productivity. The minimum requirements of minerals for beef cattle are not the required levels for maximum productivity. Critical tolerances indicate the levels of each mineral that cause poisoning or affect the level of use of other minerals.

Calcium (Ca)

Calcium is present in adequate amounts in most soils. Is a component of several primary and secondary minerals in the soil that are essentially insoluble for agricultural features. These materials are the original sources of the soluble or available forms of Ca. Moreover, Ca is present in relatively soluble forms, as a cation (positively charged Ca++) adsorbed to the soil colloidal complex. The ionic form is considered to be available to crops.

Calcium is essential for many plant functions. Some of them are:

1. Correct cell division and elongation
2. Appropriate cell wall development
3. Nitrate uptake and metabolism
4. Enzyme activity
5. Starch metabolism

Calcium is transported in the xylem via an ion exchange mechanism. It attaches to lignin molecules and exchange must occur with calcium or another similar cation (e.g. Mg++, Na+, K+, NH4+, etc.). Calcium is not very mobile in the soil, or in plant tissue, therefore a continuous supply is essential. However, there are some factors affecting Ca Availability: The Ca is found in many of the primary or secondary minerals in the soil. In this state, it is relatively insoluble. Ca is not considered a leachable nutrient. However, over hundreds of years, it will move deeper into the soil. Because of this, and the fact that many soils are derived from limestone bedrock, many soils have higher levels of Ca, and a higher pH in the subsoil. Acid soils have less Ca, and high pH soils normally have more. As the soil pH increases above pH 7.2, due to additional soil Ca, the additional "free" Ca is not adsorbed into the soil. Much of the free Ca forms nearly insoluble compounds with other elements such as phosphorus (P), thus making P less available.

Calcium is in competition with other major cations such as sodium (Na+), potassium (K+), magnesium (Mg++), Ammonium (NH4+), iron (Fe++), and aluminum (Al+++) for uptake by the plants. High K applications have been known to reduce the Ca uptake in apples, which are extremely susceptible to poor Ca uptake and translocation within the tree. High levels of soil Na will displace Ca and lead to Ca leaching. This can result in poor soil structure and possible Na toxicity to the grass. Conversely, applications of soluble Ca, typically as gypsum, are commonly used to desalinate sodic soils through the displacement principle in reverse. Soluble P is an anion, meaning it has a negative charge. Any free Ca reacts with P to form insoluble (or very slowly soluble) Ca-P compounds that are not readily available to plants. Since there is typically much more available Ca in the soil than P, this interaction nearly always results in less P availability.

As the pH of a soil decreases, more of Fe and Al become soluble and combine with Ca to form essentially insoluble compounds. High soil or plant Ca levels can inhibit Boron (B) uptake and utilization. Calcium sprays and soil applications have been effectively used to help detoxify B over-applications.

Calcium deficiency symptoms can be rather unclear since the situation often is accompanied by a low soil pH. Visible deficiency symptoms are seldom seen in agronomic crops but will typically include a failure of the new growth to develop properly. Annual grasses such as corn will have deformed emerging leaves that fail to unroll from the whorl. The new leaves are often chlorotic. Extremely acid soils can introduce an entirely new set of symptoms, often from different toxicity's and deficiencies. Calcium, for all practical purposes, is not considered to have a directly toxic effect on plants. Most of the problems caused by excess soil Ca are the result of secondary effects of high soil pH. Another problem from excess Ca may be the reduced uptake of other cation nutrients. Before toxic levels are approached in the plant, crops will often suffer deficiencies of other nutrients, such as P, K, Mg, B, Cu, Fe, or Zn.

Calcium is used in the formation and maintenance of bones and teeth. It also functions in transmission of nerve impulses and contraction of muscle tissue. A dynamic system involving Ca, P and vitamin D exists to maintain a relatively stable concentration of Ca in the blood. Calcium and P are stored in bone and mobilized into the circulatory system when dietary intake of the two minerals is adequate (Figure 8.2). Blood Ca level is not a good indicator of a dietary calcium deficiency because blood calcium is reflective of both Ca intake and Ca mobilization from bone.

Because of its importance in bone structure, deficiency of Ca in young animals leads to skeletal deformities. In older animals, fragile bones can result from extended periods of dietary Ca deficiency. Critical times to ensure that diets contain adequate Ca are during pregnancy (for proper bone growth of the fetus) and during lactation (to prevent excessive Ca mobilization from the bones of the lactating cow). Excessive mobilization of calcium from the skeletal system of the lactating cow can lead to milk fever, also known as parturient paresis or hypocalcemia. Symptoms include muscle stiffness and tremors, extreme weakness, and loss of consciousness. A common method of minimizing the risk of milk fever is to reduce Ca intake by cows for two weeks before calving. This ensures that the Ca mobilization system is functioning properly before lactation. After calving, dietary Ca is increased to meet the requirement of the lactating cow.

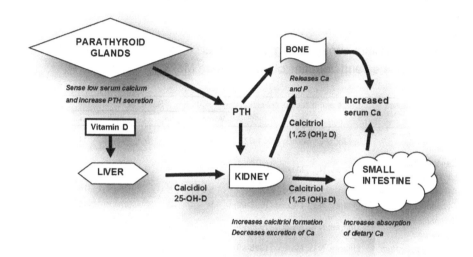

Figure 8.2. Calcium metabolism in the animal

Calcium requirements change depending on animal age and production status. Nonlactating, pregnant cows require Ca at a level of 0.18% of total dry matter intake, while the requirement for lactating cows is 0.27% of total dry matter intake. Growing and finishing cattle require 0.31% Ca for optimal growth. The maximum tolerable level of Ca is not known.

Calcium content in cultivated grasses

Seasonal Ca content in cultivated grasses that grow in northeastern Mexico ranged from 5 to 7 g kg^{-1} DM with an annual mean of 6 g kg^{-1} DM. Grasses such as *C. ciliaris* and *C. dactylon* had higher annual means compared to other grasses shown in Table 8.2. It has been reported that the average of 360 samples of tropical grasses growing around the world was 4 g kg^{-1}. This value is similar to that found in new genotypes of buffelgrass, planted under rainfed conditions without fertilization in Teran county of the state of Nuevo León, Mexico and harvested in August and November 1999 and November 2000 (Tables 8.3 and 8.4). However, when genotypes that are listed in Table 8.3 were fertilized with 120 kg ha^{-1} of urea N, surprisingly Ca content increased twice from 4 to 8 g kg^{-1} in the dry matter of grasses.

Table 8.2. Calcium content (g kg^{-1}; dry matter) in cultivated grasses collected in different dates and counties of northeastern México

Cultivated grasses	Dates and places of collection	Seasons				Annual
		Winter	Spring	Summer	Fall	mean
Cenchrus ciliaris Common	Teran, N.L., Mexico (2001-02)	4	10	8	11	8
Cenchrus ciliaris common	Linares, N.L., Mexico (1998-99)	6	4	5	7	5
Cynodon dactylon II	Marin, N.L., Mexico (1994)	5	5	7	6	6
Cynodon dactylon	Linares, N.L., Mexico (1998-99)	5	8	11	10	8
Dichanthium annulatum	Linares, N.L., Mexico (1998-99)	7	9	5	7	7
Panicum coloratum	Linares, N.L., Mexico (1998-99)	4	5	5	4	5
Rhynchelytrum repens	Teran, N.L., Mexico (2001-02)	6	8	6	6	6
Mean		5	7	7	7	6

Obtained from: Ramírez et al. (2002); Ramirez et al. (2003); Ramirez-Lozano (2003); Ramirez et al. (2004); Ramirez et al. (2005).

It has been reported that Ca content in leaves is twice as much of that in stems, so that the ruminant selective consumption of leaves may lead to their diets with higher percentages of Ca. However, with the plant maturation process it occurs a decrease in the content of Ca in the leaves compared to stems. However, the Ca in leaves of cultivated in four cultivated grasses was three times greater than that of stems (Figure 8.3). *Cynodon dactylon* had the highest content of Ca in the leaves and in all seasons; however, *C. ciliaris* had the lowest content. Intermediate values resulted in *Dichanthium annulatum* and *Panicum coloratum* (Figure 8.3).

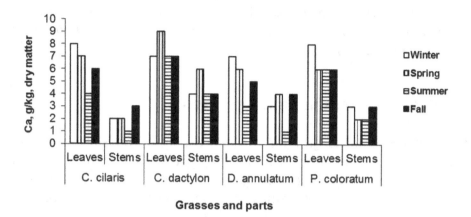

Figure 8.3. Seasonal content of Ca in leaves and stems of four cultivated grasses
collected in Linares county of the state of Nuevo Leon, Mexico in 1998 and 1999
Obtained from: Ramírez et al. (2003); Ramírez-Lozano (2003); Ramírez et al. (2005a).

Table 8.3. Calcium content (g kg^{-1}; dry matter) of the hybrid buffelgrass
Nueces y and five new genotypes of *Cenchrus ciliaris* collected in different
dates in Teran county of the state of Nuevo Leon, Mexico

Genotypes	Dates of collection			
	Aug. 1999	Nov. 1999	Nov. 2000	Jun. 2000 (Fertilized)
Cenchrus ciliaris Nueces	4	4	4	8
Cenchrus ciliaris 307622	4	4	4	8
Cenchrus ciliaris 409252	4	4	4	8
Cenchrus ciliaris 409375	4	4	4	8
Cenchrus ciliaris 409460	4	4	4	8
Cenchrus ciliaris 443	4	4	4	8
Mean	4	4	4	8

Obtained from: Garcia-Dessommes *et al.* (2003ab)

It seems that the mineral composition of grasses varies depending
on soil and plant (species), variety, part of the plant and other factors.
The Ca content in 22 species of warm-weather grasses harvested
during the rainy season in 1993 and 1995 in four geographic regions

of the Gulf Coast of Mexico, also reported variations in the Ca content among species ranging from 2.3 to 4.4 g kg^{-1} resulting *Cynodon dactylon* with the highest content and *Pennisetum purpureum* was lowest (Figure 8.4).

Calcium content in native grasses

The native grasses growing in Marin county of the state of Nuevo Leon, Mexico and harvested in 1994 (Table 8.5) had higher Ca content than those cultivated grasses listed in the Tables above. *Panicum hallii* had the highest seasonal content and *A. longiseta* was lowest. However, the native grasses that grow in Teran county of the state of Nuevo Leon, Mexico and harvested in fall of 2001 winter, spring and summer of 2002 (Table 8.6) had lower contents of Ca. *Brachiaria fasciculata*, *Digitaria insularis*, *Panicum obtusum* and *Panicum unispicatum* had higher content and *Bouteloua curtipendula*, *Chloris ciliata*, *Setaria* grisebachii and *Tridens eragrostoides* were lower.

It seems that all grasses appearing in Tables 8.2 throughout 8.6 contained levels of Ca in sufficient amounts to cover the requirements of growing beef cattle, at the beginning of lactation and adult goats (Table 8.7). Moreover, most native grasses (Tables 8.5 and 8.6) had Ca concentrations to satisfy the requirements of adult sheep in maximum production.

Magnesium (Mg)

Technically, Mg is a metallic chemical element that is vital for animal and plant life. Magnesium is one of thirteen mineral nutrients that come from soil and when dissolved in water, is absorbed through the plant's roots. Magnesium plays an important role in photosynthesis in plants. Without Mg, chlorophyll cannot capture sun energy that is needed for photosynthesis to occur. In addition, Mg is required to give leaves their green color. Magnesium in plants is located in the enzymes, in the heart of the chlorophyll molecule. Magnesium is also used by plants for the metabolism of carbohydrates and in the cell membrane stabilization. Moreover, Mg is vital to plant growth and health.

Table 8.4. Calcium content (g kg⁻¹; dry matter) of 84 new genotypes of *Cenchrus ciliaris* collected in Teran county of the state of Nuevo Leon, México in November 2000.

Genotypes	Ca, g kg⁻¹	Genotypes	Ca, g kg⁻¹	Genotypes	Ca, g kg⁻¹	Genotypes	Ca, g kg⁻¹
202513	4	409185	3	409263	4	409459	7
253261	3	409197	3	409264	2	409460	5
307622	6	409200	4	409266	2	409465	3
364428	4	409219	4	409270	4	409466	6
364439	3	409220	3	409278	3	409472	4
364445	2	409222	4	409280	3	409480	3
365654	3	409223	4	409300	4	409529	3
365702	3	409225	4	409342	3	409691	4
365704	3	409241	4	409359	5	409711	4
365713	5	409248	4	409363	4	414447	5
365728	4	409255	4	409369	3	414451	7
365731	6	409261	4	409373	4	414454	7
409142	5	409268	4	409375	6	414460	4
409151	4	409275	4	409377	6	414467	7
409154	3	409282	5	409381	5	414499	6
409155	3	409289	5	409391	7	414511	6
409157	3	409296	5	409400	3	414512	7
409162	3	409303	5	409410	6	414520	5
409164	6	409310	5	409424	8	414532	5
409165	3	409316	5	409448	5	443	6
409168	3	409258	4	409449	5	NUECES	3
Mean of all genotypes							4

Obtained from: Morales-Rodríguez *et al.*, 2005b

Magnesium is essential for many plant functions. Some of them are:

1. Photosynthesis: Mg is the central element of the chlorophyll molecule.
2. Carrier of Phosphorus in the plant
3. Magnesium is both an enzyme activator and a constituent of many enzymes
4. Sugar synthesis
5. Starch translocation

6. Plant oil and fat formation
7. Nutrient uptake control
8. Increase Iron utilization
9. Aid nitrogen fixation in legume nodules

Table 8.5. Seasonal content of Ca (g kg^{-1}; dry matter) in native grasses collected in Marin county of the state of Nuevo Leon, Mexico in 1994

| Native grasses | Seasons | | | | Annual |
	Winter	Spring	Summer	Fall	mean
Aristida longiseta	5	7	6	9	7
Bouteloua gracilis	6	12	7	13	10
Cenchrus incertus	6	10	7	11	9
Hilaria belangeri	7	10	9	10	9
Panicum hallii	9	12	11	14	12
Setaria macrostachya	5	8	8	12	8
Seasonal mean	6	10	8	12	9

Obtained from: Ramirez et al. (2004)

Magnesium deficiency in plants is common where soil is not rich in organic matter or is very light. The classic deficiency symptom is interveinal chlorosis of the lower/older leaves. However, the first symptom is generally a more pale green color that may be more pronounced in the lower/older leaves. In some plants, the leaf margins will curve upward or turn a red-brown to purple in color. Full season symptoms include preharvest leaf drop, weakened stalks, and long branched roots. Conifers will exhibit yellowing of the older needles, and in the new growth, the lower needles will go yellow before the tip needles.

Magnesium is an essential macromineral, which is required for numerous functions in the body of all mammals including dairy and beef cattle. The distribution of Mg in the body of beef and dairy cattle include nearly 70% within the bone mineral, 29% within the cells of the body and 1% in blood and extracellular fluids. The main roles of Mg in animals include:

Table 8.6. Seasonal content of Ca (g kg-1; dry matter) in native grasses collected in Teran county of the state of Nuevo Leon, Mexico in 2001 and 2002

Native grasses	Seasons				Annual
	Winter	Spring	Summer	Fall	mean
Bouteloua curtipendula	4	4	5	3	4
Bouteloua trifida	3	5	7	5	5
Brachiaria fasciculata	7	8	6	7	7
Digitaria insularis	9	7	6	7	7
Chloris ciliata	3	4	4	4	4
Leptochloa filiformis	5	5	4	6	5
Panicum hallii	6	7	6	5	6
Panicum obtusum	9	9	5	7	7
Panicum unispicatum	7	5	8	6	7
Setaria grisebachii	4	4	4	5	5
Setaria macrostachya	3	4	5	3	4
Tridens eragrostoides	4	4	5	5	4
Tridens muticus	5	5	5	5	5
Seasonal mean	5	5	5	5	5

Obtained from: Cobio-Nagao (2004)

Table 8.7. Macromineral requirements by ruminants

Elements	Beef cattle		Dairy cattle	Adult sheep	Adult goats
	Growing and finalization	Beginning of lactation			
Ca, g kg^{-1}	1.9 - 7.3	2.2 - 3.8	4.3 - 7.7	2.0 - 8.2	1.3 - 3.3
Mg, g kg^{-1}	1.0	2.0	2.0 - 2.5	1.2 - 1.8	0.8 - 2.5
K, g kg^{-1}	6.0	7.0	9.0 - 10.0	5.0 - 8.0	1.8 - 2.5
Na, g kg^{-1}	0.6 – 0.8	1.0	--	0.9 - 1.8	0.6 - 1.0
P, g kg^{-1}	1.2 - 3.4	1.6 - 2.4	2.5 - 4.8	1.6 - 3.8	1.6 - 3.8

Obtained from: McDowell (2003).

1. Bone formation
2. Cell replication
3. Hormone regulation
4. Nerve impulse transmission (stress) and muscle contraction
5. Fertility
6. Resistance to infection (immune system, vitamin E)

Grass tetany or hypomagnesemic tetany, also known as grass staggers and winter tetany, is a metabolic disease involving Mg deficiency, which can occur in such ruminant livestock as beef cattle, dairy cattle and sheep, usually after grazing on lush pastures of rapidly growing grass, especially in early spring (Figure 8.4). Low Mg intake by grazing ruminants may occur especially with some grass species early in the growing season, due to seasonally low Mg concentrations in forage dry matter. Some conserved forages are also low in Mg and may be conducive to hypomagnesemia.

High K intake relative to Ca and Mg intake may induce hypomagnesemia. A K/(Ca+Mg) charge ratio exceeding 2.2 in forages has been commonly considered a risk factor for grass tetany. Potassium fertilizer application to increase forage production may contribute to an increased K/(Ca+Mg) ratio in forage plants, not only by adding potassium to soil, but also by displacing soil-adsorbed Ca and Mg by ion exchange, contributing to increased susceptibility of Ca and Mg to leaching loss from the root zone during rainy seasons. In ruminants, high K intake results in decreased absorption of Mg from the digestive tract.

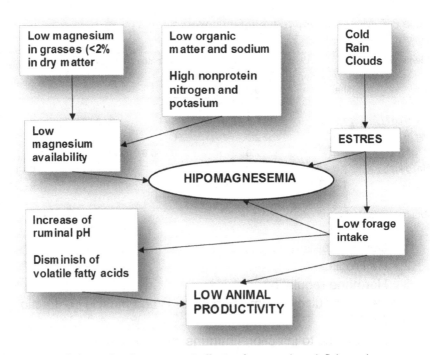

Figure 8.4. Schematic of causes and effects of magnesium deficiency in grasses

Mg content in cultivated grasses

The content of Mg in tropical grasses from around the world is very variable. It has been reported that the level of Mg in 280 tropical grasses ranged from 0.4 to 9.0 g kg⁻¹ dry matter basis, with an average of 3.6 g kg⁻¹. Lower content of Mg was found in 22 tropical grasses growing in different regions of the Gulf Coast of Mexico was assessed with a range of 0.5 to 2.1 g kg⁻¹ with a mean of 1.5 g kg⁻¹ dry basis. Cultivated grasses listed in Tables 6.7, 6.8 and 6.9 also had low Mg concentrations. It is possible that differences in Mg content are due grasses were collected for analyses in different places and different periods of time; besides the dissimilarities between species.

Table 8.8. Magnesium content (g kg⁻¹) in cultivated grasses collected in different dates and counties of the state of Nuevo Leon, Mexico

Grasses	Dates and places of collection	Seasons				Annual
		Winter	Spring	Summer	Fall	mean
Cenchrus ciliaris, common	Teran, N.L., México (2001-02)	1	2	1	2	2
Cenchrus ciliaris, common	Linares, N.L., México (1998-99)	2	1	2	1	2
Cynodon dactylon II	Marin, N.L., México (1994)	1	1	2	1	1
Cynodon dactylon	Linares, N.L., México (1998-99)	2	2	2	2	2
Dichanthium annulatum	Linares, N.L., México (1998-99)	1	1	1	1	1
Panicum coloratum	Linares, N.L., México (1998-99)	1	1	2	1	1
Rhynchelytrum repens	Teran, N.L., México (2001-02)	2	2	3	2	2
Mean		1	1	2	1	1

Obtained from: Ramirez *et al.* (2002); Ramirez *et al.* (2003); Ramirez-Lozano (2003); Ramirez *et al.* (2004); Ramirez *et al.* (2005).

It seems that, except in summer, dairy cattle and adult sheep consuming the grasses listed in Table 6.8 may suffer Mg deficiency, which could be manifested by tetanus hypomagnesemia, it is possible because most plants had low marginally concentrations of Mg to meet the requirements of livestock (Table 8.7). Individually, grasses such as *Cynodon dactylon* and *Rhynchelytrum repens* contain levels of Mg (Table 8.8) to meet the needs of adult ruminants.

When the genotypes that are listed in Table 8.9 were fertilized with urea at 120 kg ha⁻¹, Mg content was significantly increased when compared with the same genotypes but unfertilized. Furthermore, it was reported that the Mg increase in fertilized grasses with N as urea, presumably because most of the N is obtained by the plant in the form of nitrate.

Table 8.9. Magnesium content (g kg^{-1}; dry matter) of the hybrid buffelgrass Nueces and five new genotypes of Cenchrus ciliaris collected in different dates inn Teran, Nuevo Leon, Mexico

Grasses	Place of collection	Dates of collection			
		Aug. 1999	Nov. 1999	Nov. 2000	Jun. 2000 (Fertilized)
Cenchrus ciliaris Nueces	Teran, N.L., Mexico	5	5	5	26
Cenchrus ciliaris 307622	Teran, N.L., Mexico	3	4	4	17
Cenchrus ciliaris 409252	Teran, N.L., Mexico	4	4	4	19
Cenchrus ciliaris 409375	Teran, N.L., Mexico	4	4	4	23
Cenchrus ciliaris 409460	Teran, N.L., Mexico	5	4	4	22
Cenchrus ciliaris 443	Teran, N.L., Mexico	4	5	4	19
Mean		4	4	4	21

Obtained from: Garcia-Dessommes et al. (2003ab)

A similar Mg trend that had the genotypes reported in Table 8.9 was observed in those listed in Table 8.10. The Mg content varied from 2 to 5 g kg^{-1} with a mean value of 3 g kg^{-1}. Growing cattle requires 1.0 g kg^{-1} of Mg in the diet. Therefore, ruminants fed the 84 genotypes shown in Table 6.10, may not suffer Mg deficiency. In tropical grasses, unlike Ca, Mg content in leaves and stems is similar. The same pattern from leaves and stems was found in the grasses C. ciliaris, C. dactylon, D. annulatum and P. coloratum (Figure 8.5) collected in different seasons, under rainfed conditions in Linares county of the state of Nuevo Leon, Mexico in 1998 and 1999, containing 2.3 and 2.4 g Mg kg^{-1} DM in leaves and stems, respectively.

Magnesium content in native grasses

Even though the native grasses that grow in Marin y Teran counties of the state of Nuevo Leon, Mexico (Tables 8.11 and 8.12, respectively) containing Mg levels below those of cultivated grasses, the growing beef cattle fed with native grasses, might not have Mg deficiency (Table 8.7). However, grasses such as A. longiseta (Table 8.10.), Bouteloua trifida, and B. curtipendula (Table 8.12), in all seasons, they contain insufficient amounts of Mg to meet the demands of growing beef cattle. In summer and autumn, which are the rainy seasons in this region of northeastern Mexico, the percentage of Mg was higher compared to other seasons.

Table 8.10. Magnesium content (g kg^{-1}; dry matter) of 84 new genotypes of *Cenchrus ciliaris* collected in Teran county of the state of Nuevo Leon, Mexico in November 2000

Genotypes	Mg, g kg^{-1}	Genotypes	Mg, g kg^{-1}	Genotypes	Mg, g kg^{-1}	Genotypes	Mg, g kg^{-1}
202513	2	409185	2	409263	3	409459	3
253261	2	409197	2	409264	3	409460	3
307622	4	409200	2	409266	3	409465	3
364428	2	409219	3	409267	3	409466	3
364439	3	409220	2	409269	3	409472	3
364445	2	409222	3	409270	3	409480	3
365654	2	409223	3	409272	3	409529	2
365702	2	409225	2	409273	4	409691	3
365704	3	409227	2	409275	4	409711	3
365713	2	409228	3	409276	4	414447	3
365728	3	409229	2	409278	4	414451	3
365731	3	409230	2	409279	4	414454	3
409142	3	409232	2	409281	4	414460	3
409151	3	409234	3	409282	4	414467	4
409154	2	409235	3	409284	4	414499	3
409155	2	409238	3	409285	4	414511	3
409157	2	409240	3	409287	5	414512	3
409162	2	409242	3	409288	5	414520	4
409164	3	409252	3	409290	5	414532	3
409165	3	409254	3	409291	5	443	2
409168	2	409258	3	409293	5	Nueces	3
Mean of all grasses							3

Obtained from: Morales-Rodríguez *et al.*, 2005b

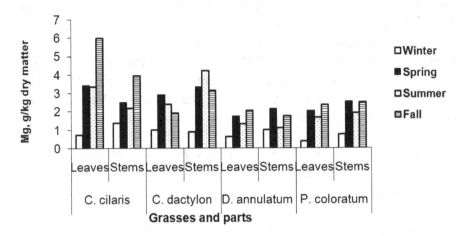

Figure 8.5. Seasonal Mg content (g kg^{-1} dry matter) in leaves
and stems of four cultivated grasses collected in Linares county
of the state of Nuevo Leon, Mexico in 1998 y 1999.
Obtained from: Ramirez *et al.* (2003); Ramirez-Lozano (2003); Ramirez *et al.* (2005a).

Potassium (K)

Potassium is vital to many plant processes. Potassium is considered
second only to nitrogen, when it comes to nutrients needed by plants,
and is commonly considered as the "quality nutrient." A review of its
role involves understanding the basic biochemical and physiological
systems of plants. While K does not become a part of the chemical
structure of plants, it plays many important regulatory roles in
development such as:

1. Increases root growth and improves drought resistance
2. Activates many enzyme systems
3. Maintains turgor; reduces water loss and wilting
4. Aids in photosynthesis and food formation
5. Reduces respiration, preventing energy losses
6. Enhances translocation of sugars and starch
7. Produces grain rich in starch
8. Increases protein content of plants
9. Builds cellulose and reduces lodging
10. Helps retard crop diseases

Table 8.11. Magnesium content (g kg-1 dry matter) in native grasses collected in Marin county of the state of Nuevo Leon, Mexico in 1994

Native grasses	Seasons				Annual
	Winter	Spring	Summer	Fall	Mean
Aristida longiseta	0.8	0.9	0.6	1.0	0.8
Bouteloua gracilis	0.6	1.1	1.1	1.5	1.1
Cenchrus incertus	1.0	2.3	2.0	2.7	2.0
Hilaria belangeri	0.5	1.3	1.0	1.6	1.1
Panicum hallii	1.3	1.8	1.7	2.2	1.8
Setaria macrostachya	1.1	1.2	1.0	2.1	1.4
Mean	0.9	1.4	1.2	1.9	1.4

Obtained from: Ramirez *et al.* (2004)

Table 8.12. Magnesium content (g kg-1 dry matter) in native grasses collected in Teran county of the state of Nuevo Leon, Mexico

Native grasses	Seasons				Annual
	Winter	Spring	Summer	Fall	Mean
Bouteloua curtipendula	0.7	0.9	0.7	0.7	0.8
Bouteloua trifida	0.6	0.9	0.1	0.8	0.6
Brachiaria fasciculata	2.2	1.1	3.1	3.7	2.5
Digitaria insulares	1.8	1.8	1.8	1.9	1.8
Chloris ciliata	0.7	2.1	1.1	1.1	1.3
Leptochloa filiformis	1.6	0.8	1.6	2.1	1.5
Panicum hallii	1.3	0.7	2.8	3.2	2.0
Panicum obtusum	2.1	1.5	1.6	1.6	1.7
Panicum unispicatum	1.5	2.5	3.4	2.0	2.4
Setaria grisebachii	2.0	1.1	2.0	2.5	1.9
Setaria macrostachya	1.2	1.2	2.8	1.3	1.6
Tridens eragrostoides	1.3	1.2	1.2	1.3	1.3
Tridens muticus	0.7	2.2	1.3	1.0	1.3
Mean	1.4	1.4	1.8	1.8	1.6

Obtained from: Cobio-Nagao (2004)

Potassium deficiency is a plant disorder that is most common on light, sandy soils, because K ions (K+) are highly soluble and will easily leach from soils without colloids. Potassium deficiency is also common in chalky or peaty soils with a low clay content. It is also found on heavy clays with a poor structure. Distinctive symptoms of K deficiency in plants include brown scorching and curling of leaf tips as well as chlorosis (yellowing) between leaf veins. Purple spots may also appear on the leaf undersides. Plant growth, root development, and seed and fruit development are usually reduced in K-deficient plants. Often, K deficiency symptoms first appear on older (lower) leaves because K is a mobile nutrient, meaning that a plant can allocate K to younger leaves when it is K deficient. Deficient plants may be more prone to frost damage and disease, and their symptoms can often be confused with wind scorch or drought.

Potassium (K) is essential for human and animal life. Potassium is involved in many body functions and is required for proper muscle development. Adequate K is also important for good heart function. The recommended daily allowance (RDA) of K varies depending on species, stage of growth, and level of other dietary minerals. Potassium functions in the intracellular fluids the same as Na does in the extracellular fluids (Figure 8.6). The major functions of K in the animal body are to:

1. Maintain water balance
2. Maintain osmotic pressure
3. Maintain acid-base balance
4. Activate enzymes
5. Help metabolize carbohydrates and proteins
6. Regulate neuromuscular activity (along with Ca)
7. Help regulate heartbeat.

Potassium deficiency are manifested by depressed growth, muscular weakness, stiffness, decreased feed intake, intracellular acidosis, nervous disorders, reduced heart rate, and abnormal electrocardiograms. The first sign of K deficiency is reduced feed intake. Many of the other signs stem from reduced feed intake. Potassium must be supplied in the daily ration because it is a mobile nutrient and there are not any appreciable reserves.

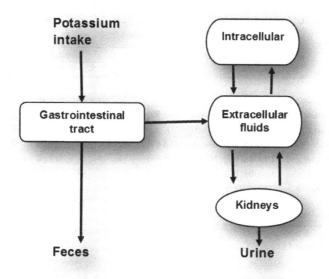

Figure 8.6. Homeostatic balance of potassium in the animal body

Potassium content in cultivated grasses

Potassium is the most abundant element in plants as Ca is the in animals. Growing beef cattle require 6 g kg^{-1} of K in their diet for not suffer deficiency. All grasses listed in Tables 8.13, 8.14 and 8.15 contain K levels to support the metabolic demands of K for growing ruminants. However, a much higher range (1.0-22.5 g kg^{-1} dry matter) was reported in 22 tropical grasses growing and collected in several regions of the Gulf Coast of Mexico.

It appeared that, cultivated grasses had greater content of K in summer when the highest rainfall occurred. However, at the end of fall and in winter, when maturity occurs, it had little effect on K content of grasses listed in Table 8.13. It was also found that changes in the concentrations of K as maturity advances, grasses, are less consistent than the changes in the contents of N, P or S. However, there is a wide difference in the content of K among grass species (Table 8.13) and the date of collection within the same species or whether they are fertilized with N as urea (Table 8.14).

Table 8.13 Potassium content (g kg⁻¹; dry matter) in cultivated grasses collected in different counties of the state of Nuevo Leon, Mexico

Grasses	Place and date of collection	Seasons				Annual
		Winter	Spring	Summer	Fall	mean
Cenchrus ciliaris Common	Teran, N.L., Mexico (2001-02)	16	29	16	24	21
Cenchrus ciliaris Common	Linares, N.L., Mexico (1998-99)	9	8	16	10	11
Cynodon dactylon II	Marin, N.L., Mexico (1994)	27	19	32	29	27
Cynodon dactylon	Linares, N.L., Mexico (1998-99)	22	23	26	22	23
Dichanthium annulatum	Linares, N.L., Mexico (1998-99)	9	11	8	9	9
Panicum coloratum	Linares, N.L., Mexico (1998-99)	26	19	30	24	25
Rhynchelytrum repens	Teran, N.L., Mexico (2001-02)	7	7	12	8	9
Promedio		17	17	20	18	18

Obtained from: Ramirez *et al.* (2002); Ramirez *et al.* (2003); Ramirez-Lozano (2003); Ramirez *et al.* (2004); Ramirez *et al.* (2005).

Table 8.14. Potassium content (g kg⁻¹; dry matter) of the hybrid buffelgrass Nueces and five new genotypes of *Cenchrus ciliaris* collected in different dates of Teran county of the state of Nuevo Leon, Mexico

Líneas de buffel	Fechas de colecta			
	Aug.1999	Nov. 1999	Nov. 2000	Jun. 2000 (Fertilized)
Cenchrus ciliaris Nueces	22	21	24	30
Cenchrus ciliaris 307622	21	18	29	35
Cenchrus ciliaris 409252	15	12	24	29
Cenchrus ciliaris 409375	21	15	21	47
Cenchrus ciliaris 409460	17	14	25	30
Cenchrus ciliaris 443	16	15	24	35
Mean	19	16	25	34

Obtained from: Garcia-Dessommes *et al.* (2003ab)

Potassium content in grasses also varies between genotypes of the same species. Table 8.15 show the K content in 84 new genotypes of buffelgrass that were planted under rainfed conditions in the Teran county of the state of Nuevo Leon, Mexico. Even they belong to the same species, they varied in a range from 9 to 32 g kg⁻¹ dry basis with a mean of 20 g kg⁻¹ dry basis.

Table 8.15. Potassium content (g kg^{-1}; dry matter) of 84 new genotypes of *Cenchrus ciliaris* collected in different dates of Teran county of the state of Nuevo Leon, in November 2000.

Genotypes	K, g kg^{-1}	Genotypes	K, g kg^{-1}	Genotypes	K, g kg^{-1}	Genotypes	K, g kg^{-1}
202513	23	409185	24	409263	24	409459	16
253261	22	409197	22	409264	21	409460	20
307622	19	409200	25	409266	21	409465	19
364428	18	409219	24	409267	22	409466	14
364439	23	409220	22	409269	25	409472	15
364445	24	409222	22	409270	23	409480	17
365654	21	409223	25	409272	23	409529	11
365702	21	409225	21	409273	19	409691	15
365704	29	409227	18	409275	27	409711	17
365713	22	409228	23	409276	15	414447	14
365728	26	409229	25	409278	22	414451	14
365731	32	409230	25	409279	23	414454	13
409142	31	409232	24	409281	11	414460	21
409151	20	409234	24	409282	21	414467	15
409154	21	409235	26	409284	20	414499	12
409155	21	409238	23	409285	21	414511	17
409157	22	409240	24	409287	17	414512	17
409162	23	409242	17	409288	19	414520	18
409164	19	409252	10	409290	16	414532	18
409165	24	409254	22	409291	17	443	9
409168	24	409258	23	409293	17	Nueces	25
Mean of all genotypes							20

Obtained from: Morales-Rodríguez *et al.*, 2005

It appears that in general, grasses such as *C. ciliaris, C. dactylon, D. annulatum* and *P. coloratum* (Figure 8.7, had more K in stems than in leaves. However, as maturation proceeds, this trend inclines to be lower. In winter and spring (dry seasons) K content was lower than in other seasons; however, K in stems is still higher than the leaves. The differences between the parties were very small.

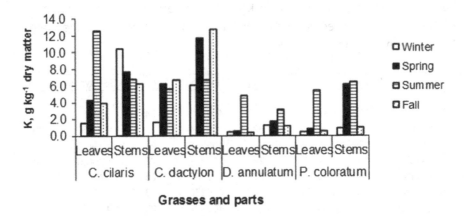

Figure 8.7. Seasonal content of K (g kg^{-1}; dry matter) in leaves
and stems of four cultivated grasses collected in Linares county
of the state of Nuevo Leon, Mexico in 1998 and 1999.
Obtained from: Ramirez et al. (2003); Ramirez-Lozano (2003); Ramirez et al. (2005a).

Variations in the content of K between species and seasons also
reported in the native grasses that grow in the counties of Teran
and Marin of the state of Nuevo Leon, Mexico (Tables 8.17 and
8.18). *Bouteloua curtipendula, B. trifida* and *Tridens muticus* had
levels throughout the year, below the requirements of growing cattle.
However, the rest of the grasses had K levels to sustain metabolic
activities of growing beef cattle. In general, during the wet seasons
(summer and fall) grasses had greater content of K compared to the
dry seasons (winter and spring).

Sodium

Sodium is the sixth most abundant element in the Earth's crust, and
exists in numerous minerals such as feldspars, sodalite and rock salt
(NaCl). Many salts of sodium are highly water-soluble. Sodium ions
have been leached by the action of water so that sodium and chlorine
(Cl) are the most common dissolved elements by weight in the Earth's
bodies of oceanic water.

In C4 plants, Na is an element that helps in metabolism, specifically
in regeneration of phosphoenolpyruvate (biosynthesis of various

aromatic compounds, and in carbon fixation) and synthesis of chlorophyll. In others, it substitutes for potassium in several roles, such as maintaining turgor pressure and aiding in the opening and closing of stomata.

Table 8.16. Seasonal K content (g kg-1 dry matter) in native grasses collected in Marin county of the state of Nuevo Leon, Mexico in 1994

| Native grasses | Seasons | | | | Annual |
	Winter	Spring	Summer	Fall	mean
Aristida longiseta	2	4	2	4	3
Bouteloua gracilis	4	6	4	7	5
Cenchrus incertus	19	24	24	28	24
Hilaria belangeri	4	7	6	7	4
Panicum hallii	11	16	14	15	14
Setaria macrostachya	8	14	10	18	13
Seasonal mean	8	12	10	13	11

Obtained from: Ramirez et al. (2004)

Excess sodium in the soil limits the uptake of water due to decreased water potential, which may result in wilting; similar concentrations in the cytoplasm can lead to enzyme inhibition that may causes necrosis and chlorosis. To avoid these problems, plants developed mechanisms that limit sodium uptake by roots, store them in cell vacuoles, and control them over long distances; excess sodium may also be stored in old plant tissue, limiting the damage to new growth.

Sodium plays a key role in the mechanism by which cells move nutrients back and forth across their borders. It is necessary for transmission of nerve impulses those signals responsible for contraction of skeletal, heart and digestive tract muscles (Figure 8.8). Sodium is a major component of saliva, too, and helps buffer acid during ruminal fermentation. Because salt affects how the body functions at a cellular level, the most common and most costly result of salt deficiency might be reduced performance. Every aspect of performance is affected. Growth, fertility and reproduction, and milk production decline. Cattle simply do not perform to their genetic potential.

Table 8.17. Seasonal K content (g kg-1 dry matter) in native grasses collected in Teran of the state of Nuevo Leon, Mexico in 2001 and 2002

Native grasses	Seasons				Annual mean
	Winter	Spring	Summer	Fall	
Bouteloua curtipendula	3	7	6	4	5
Bouteloua trifida	2	4	4	2	3
Brachiaria fasciculata	6	2	33	18	15
Digitaria insularis	10	7	21	15	13
Chloris ciliata	7	16	19	9	13
Leptochloa filiformis	4	3	10	9	7
Panicum hallii	5	2	19	13	10
Panicum obtusum	13	11	27	17	17
Panicum unispicatum	7	21	16	13	14
Setaria grisebachii	8	7	33	19	17
Setaria macrostachya	19	11	40	29	25
Tridens eragrostoides	14	19	11	13	14
Tridens muticus	2	6	7	4	5
Seasonal means	8	9	19	13	12

Obtained from: Cobio-Nagao (2004)

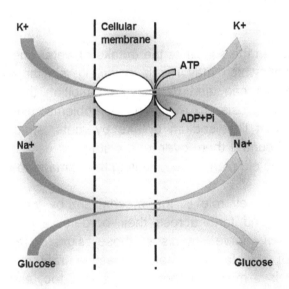

Figure 8.8. Sodium pumps vital for maintaining the electrochemical differences between the cell membranes and glucose consumption by the action of Na + and Pi ATP-dependent

Deficiency of Na is more likely to occur during the following circumstances:

1. During lactation due to deposition of Na in milk
2. In fast growing animals
3. Under tropical conditions or semiarid hot climate
4. In cattle, consuming grasses heavily fertilized with K, which reduces levels of Na.

Even after prolonged deficiency, Na levels in the milk remain high. The first sign of Na and Cl deficiency is craving for salt, shown by constant licking of wood, dirt and sweat of other animals, and water consumption. A prolonged deficiency causes loss of appetite, poor appearance, low milk production, weight loss and reduced growth. The most pronounced signs of deficiency include incoordination, body tremors, weakness and loss of heart rhythm, which can lead to death.

The Na and Cl ions are absorbed in the gastrointestinal tract of ruminants. The needs of Na and Cl in ruminants are 0.1- 0.2% of dry matter, respectively. The Na and Cl in lesser extent, not always are found in normal rations in sufficient quantities. Therefore, it is normal rations to supplement with common salt. The Cl excess in the diet can cause acidosis and alkalosis the excess of sodium.

Sodium content in cultivated grasses

The content of Na in tropical grasses is highly variable between species. It has been reported that in 192 samples of tropical grasses grown in various parts of the world ranged from 0.1 to 18.0 g kg^{-1} dry matter, with a mean of 2.6 g kg^{-1}. 52% of the samples contained 1.0 g kg^{-1} Na, and another 18% were between 4.0 and 8.0 g kg^{-1}. Large differences in the content of Na between grasses may be because some species are Na accumulators and others are not. Growing beef cattle requires about 0.8 g kg^{-1} in of diet dry matter to meet their needs of Na. Except for some native grasses that grow in northeastern, Mexico, most of the grasses listed in Tables 8.18, 8.19, 8.20 and 8.21 were deficient in Na to support the needs of ruminants growth.

Table 8.18. Seasonal sodium content (g kg^{-1}; dry matter) in cultivated grasses collected in different dates and counties of the state of Nuevo Leon, Mexico

Cultivated grasses	Dates and places of collection	Seasons				Annual
		Winter	Spring	Summer	Fall	mean
Cenchrus ciliaris Common	Teran, N.L., México (2001-02)	0.2	0.2	0.4	0.6	0.4
Cenchrus ciliaris Common	Linares, N.L., México (1998-99)	0.6	1.0	0.7	1.4	0.9
Cynodon dactylon II	Marin, N.L., México (1994)	1.0	0.4	1.0	0.6	0.8
Cynodon dactylon	Linares, N.L., México (1998-99)	0.4	2.0	1.0	1.0	1.1
Dichanthium annulatum	Linares, N.L., México (1998-99)	1.1	1.0	1.3	1.2	1.1
Panicum coloratum	Linares, N.L., México (1998-99)	0.4	0.3	1.0	1.0	0.7
Rhynchelytrum repens	Teran, N.L., México (2001-02)	0.1	0.2	0.1	0.2	0.2
Mean		0.4	0.7	0.7	0.8	0.6

Obtained from: Ramirez et al. (2002); Ramirez et al. (2003); Ramirez-Lozano (2003); Ramirez et al. (2004); Ramirez et al. (2005).

Na deficiency does not cause adverse effects in ruminants, so the real needs could be considerably lower than recommended. In addition, cattle have large reserves of Na and retain very effectively when they receive Na deficient diets. Dry cows must spend at least six months with a diet deficient pastures Na to acquire the Na deficiency. However, in milking cows, a drop may occur in milk production at two months of receiving a diet low in Na, since the cows are not able to reduce the amount of Na contained in the milk.

Table 8.19. Sodium content (g kg^{-1}; dry matter) of the hybrid buffelgrass Nueces and five new genotypes of Cenchrus ciliaris collected in different dates in Teran county of the state of Nuevo Leon, Mexico

Genotypes	Dates of collection			
	Aug. 1999	Nov. 1999	Nov. 2000	Jun. 2000 (Fertilized)
Cenchrus ciliaris Nueces	0.11	0.11	0.12	0.12
Cenchrus ciliaris 307622	0.11	0.11	0.17	0.13
Cenchrus ciliaris 409252	0.12	0.12	0.12	0.10
Cenchrus ciliaris 409375	0.14	0.1	0.12	0.11
Cenchrus ciliaris 409460	0.11	0.1	0.24	0.12
Cenchrus ciliaris 443	0.10	0.1	0.11	0.11
Mean	0.12	0.11	0.15	0.12

Obtained from: Garcia-Dessommes et al. (2003ab)

It appears that the application of nitrogen fertilizer in the form of urea has no effect on the content of Na in the grasses. Similar results have been reported when comparing new genotypes of *Cenchrus ciliaris* (Table 8.18) that were planted at different dates in the Teran county of the state of Nuevo Leon, Mexico and were fertilized with 120 kg ha[-1] urea N and irrigated.

Cultivated grasses such as *C. ciliaris, C. dactylon, D. annulatum* and *P. coloratum* (Figure 8.9) contain more Na in the leaves than in the stems. However, as the maturity progresses, Na content in stems and leaves remained similar. In fact, the ruminant consuming, either alone leaves or single stems of these four grasses, might have not develop a syndrome of deficiency because all grasses and in all seasons have Na in sufficient amounts to meet the demands of ruminants.

Table 8.20. Sodium content (g kg[-1]; dry matter) of 84 new genotypes of *Cenchrus ciliaris* collected in Teran county of the state of Nuevo Leon, Mexico in November 2000

Genotypes	Na, g kg[-1]	Genotypes	Na, g kg[-1]	Genotypes	Na, g kg[-1]	Genotypes	Na, g kg[-1]
202513	0.1	409185	0.3	409263	0.3	409459	0.1
253261	0.1	409197	0.2	409264	0.2	409460	0.1
307622	0.2	409200	0.1	409266	0.1	409465	0.2
364428	0.2	409219	0.2	409267	0.1	409466	0.1
364439	0.3	409220	0.2	409269	0.1	409472	0.1
364445	0.3	409222	0.1	409270	0.1	409480	0.1
365654	0.3	409223	0.1	409272	0.1	409529	0.1
365702	0.2	409225	0.2	409273	0.1	409691	0.1
365704	0.4	409227	0.2	409275	0.2	409711	0.1
365713	0.3	409228	0.1	409276	0.1	414447	0.1
365728	0.2	409229	0.3	409278	0.2	414451	0.1
365731	0.3	409230	0.1	409279	0.2	414454	0.1
409142	0.1	409232	0.4	409281	0.1	414460	0.3
409151	0.3	409234	0.2	409282	0.1	414467	0.1
409154	0.3	409235	0.2	409284	0.1	414499	0.2
409155	0.2	409238	0.1	409285	0.1	414511	0.1
409157	0.2	409240	0.2	409287	0.2	414512	0.1
409162	0.1	409242	0.2	409288	0.1	414520	0.1
409164	0.1	409252	0.3	409290	0.2	414532	0.1
409165	0.3	409254	0.2	409291	0.1	443	0.1
409168	0.1	409258	0.2	409293	0.2	NUECES	0.3
Mean of all genotypes							0.2

Obtained from: Morales-Rodriguez *et al.*, 2005b

Sodium content of native grasses

All native grasses growing in Marin and Teran counties of the state of Nuevo Leon, Mexico (Tables 8.20 and 8.21) were Na deficient to meet the needs (0.8 g kg⁻¹ in the diet dry matter) of growing ruminants. Conversely, it has been reported that native grasses such as *Aristida longiseta* (2.2), *Bouteloua trifida* (1.3), *Hilaria mutica* (1.8) and *Setaria macrostachya* (1.5 g kg⁻¹), growing in different regions of the state of Nuevo Leon, Mexico, contained levels of Na in sustainable quantities for growing ruminants.

Table 8.22. Seasonal sodium content (g kg-1 dry matter) in native grasses collected in Marin county of the state of Nuevo Leon, Mexico in 1994

Native grasses	Seasons				Annual
	Winter	Spring	Summer	Fall	mean
Aristida longiseta	0.5	0.7	0.4	1.0	0.7
Bouteloua gracilis	0.6	1.0	0.5	1.1	0.8
Cenchrus incertus	0.5	0.7	0.4	1.3	0.7
Hilaria belangeri	0.4	0.7	0.5	0.8	0.6
Panicum hallii	0.4	0.9	0.5	1.2	0.8
Setaria macrostachya	0.3	0.7	0.5	0.9	0.6
Promedio estacional	0.5	0.8	0.5	1.1	0.7

Obtained from: Ramirez *et al.* (2004)

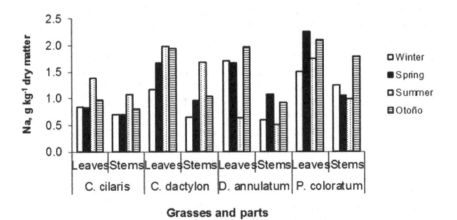

Figure 8.9. Seasonal content of Na (g kg-1 dry matter) in leaves and stems of four Cultivated grasses and collected in Linares, Nuevo Leon, México in 1998 and 1999 Obtained from: Ramirez et al. (2003); Ramirez-Lozano (2003); Ramirez et al. (2005a).

Table 8.22. Seasonal sodium content (g kg-1 dry matter) in native grasses collected in Teran county of the state of Nuevo Leon, Mexico in 2001 and 2002

| Native grasses | Seasons | | | | Annual |
	Winter	Spring	Summer	Fall	mean
Bouteloua curtipendula	0.1	0.1	0.2	0.2	0.2
Bouteloua trifida	0.2	0.1	0.2	0.2	0.2
Brachiaria fasciculata	0.1	0.2	0.4	0.2	0.2
Digitaria insularis	0.4	0.2	0.5	0.4	0.4
Chloris ciliata	0.2	0.2	0.2	0.2	0.2
Leptochloa filiformis	0.1	0.1	0.3	0.3	0.2
Panicum hallii	0.1	0.3	0.3	0.2	0.2
Panicum obtusum	0.3	0.1	0.2	0.3	0.2
Panicum unispicatum	0.2	0.4	0.2	0.2	0.3
Setaria grisebachii	0.1	0.2	0.2	0.1	0.2
Setaria macrostachya	0.3	0.2	1.1	0.4	0.5
Tridens eragrostoides	0.2	0.1	0.2	0.3	0.2
Tridens muticus	0.2	0.1	0.1	0.2	0.2
Seasonal means	0.2	0.2	0.3	0.2	0.2

Obtained from: Cobio-Nagao (2004)

Phosphorous

The primary biological importance of phosphates is as a component of nucleotides, which serve as energy storage within cells (ATP) or when linked together, form the nucleic acids DNA and RNA. The double helix of our DNA is only possible because of the phosphate ester bridge that binds the helix. Besides making biomolecules, phosphorus is also found in bone and the enamel of mammalian teeth, whose strength is derived from calcium phosphate in the form of hydroxyapatite. It is also found in the exoskeleton of insects, and phospholipids (found in all biological membranes). It also functions as a buffering agent in maintaining acid base homeostasis in the human body.

Phosphorus is one of 17 nutrients essential for plant growth. Its functions cannot be performed by any other nutrient, and an adequate supply of P is required for optimum growth and reproduction. Phosphorus is classified as a major nutrient, meaning that it is frequently deficient for crop production and is required by crops in relatively large amounts. The total P concentration in agricultural crops generally varies from 0.1 to 0.5%. The most important chemical reaction in nature is photosynthesis. It utilizes light energy in the presence of chlorophyll to combine carbon dioxide and water into simple sugars, with the energy being captured in ATP. The ATP is then available as an energy source for the many other reactions that occur within the plant, and the sugars are used as building blocks to produce other cell structural and storage components.

When P is deficient, the most outstanding effects are a reduction in leaf expansion and leaf surface area, as well as the number of leaves. Shoot growth is more affected than root growth, which leads to a decrease in the shoot:root dry weight ratio. Nonetheless, root growth is also reduced by P deficiency, leading to less root mass to reach water and nutrients. Generally, inadequate P slows the processes of carbohydrate utilization, while carbohydrate production through photosynthesis continues. This results in a buildup of carbohydrates and the development of a dark green leaf color. In some plants, P-deficient leaves develop a purple color, tomatoes and corn being two examples.

Since P is readily mobilized in the plant, when a deficiency occurs the P is translocated from older tissues to active meristematic tissues, resulting in foliar deficiency symptoms appearing on the older (lower) portion of the plant (Figure 8.11). However, such symptoms of P deficiency are seldom observed in the field...other than loss of yield. Other effects of P deficiency on plant growth include delayed maturity, reduced quality of forage, fruit, vegetable, and grain crops, and decreased disease resistance.

Phosphorus (P) is an essential nutrient for all animals. Deficiency of P is the most widespread of all the mineral deficiencies affecting livestock. Phosphorus must be balanced in the animal diet with adequate calcium (Ca) and vitamin D for growth, reproduction, gestation, and lactation. Young and growing ruminants require relatively more P than do mature ones. Gestating and lactating ruminants need more P than other classes of mature animals. Specific

P requirements for maintenance, growth, lactation, and pregnancy depend on many factors.

Recommendations in National Research Council (NRC) publications are based on complex models that consider body size, breed, milk production levels, and environmental conditions. For dairy cattle, the Ca:P ratio should be at least 2.4:1 for cows when lactating, but should be less than 1.6:1 for dry cows to minimize Ca intake during that period. For beef cows and feedlot cattle, the ratio is not so critical, although normally it would not be allowed to exceed 4:1.

Deficiency of P is the most widespread and economically important of all the mineral deficiencies affecting grazing livestock. Thus, supplemental dietary P is needed under most practical feeding situations. On grazed pasture, where soils are low in P, fertilizing with P can reduce risk of grass tetany. Phosphorus enhances reproductive performance at several stages in the reproductive cycle. Moreover, some tests showed that P increased rebreeding efficiency for beef cows. Irregular estrus periods have been associated with moderate P deficiency, infertility with marginal P levels, and anestrus with low P levels. In other experiments, 64% of the control cows produced a calf on range alone compared to 85% of the cows on range plus P supplement. In addition, P supplementation has been shown to increase fertility, calving rates, calf growth rates, and, when applied to pastures, carrying capacity.

Rumen microbes require P and obtain it from the P released during fermentation of feed in the rumen, and from salivary. It has been suggested the maximum P requirement for ruminal microbes is satisfied when the diet contains 4 g P kg^{-1} digestible organic matter. This is equivalent to <0.30% dietary P. This is probably a high estimate, however, because in vitro rumen studies suggest that P concentrations in ruminal fluid of 0.7 to 2.6 mM are adequate to support maximum microbial growth, and these concentrations can be achieved with diets containing as low as 0.12%. This depends upon how much P is contributed by saliva. It also has been suggested that available P from dietary sources and saliva be at least 5 g kg^{-1} digestible organic matter. If P is fed to meet the requirement of the animals, then there will be sufficient P from salivary and feed sources to meet the requirement of rumen microbes.

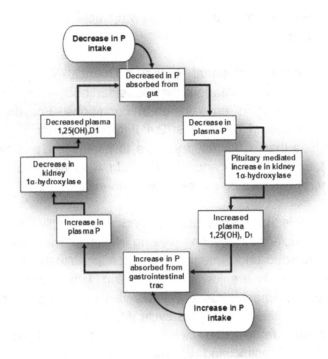

Figure 8.10. Mechanism of adaptation to alterations in dietary phosphorus

Phosphorous content in cultivated grasses

The P is the mineral most commonly deficient in grasses consumed by grazing cattle. This is especially true in tropical and subtropical areas, and most of Latin America. Under grazing conditions, either in rangeland or pastures without fertilization, phosphorus levels of grasses are well below the requirements of the ruminant. Mature forages usually contain less than 1.5 g P kg^{-1} dry matter, while the requirements of beef cattle are usually higher than 2.0 g P kg^{-1} dry matter.

The P is also deficient in growing grasses grown in semi-arid areas of northeastern Mexico. With the exception of grass *Cynodon dactylon* (Cross II) all cultivated pastures in Tables 8.22, 8.23, 8.24 P contain insufficient levels to meet the requirements of growing ruminants. *Rhynchelytrum repens* had the lowest value. Similar trend was observed in native grasses listed in Tables 8.25, 8.26 and 8.27.

Table 8.23. Seasonal phosphorous content (g kg⁻¹, dry matter) in cultivated grasses collected in different dates and counties of the state of Nuevo Leon, México

Cultivated grasses	Places and dares of collection	Seasons				Annual
		Winter	Spring	Summer	Fall	mean
Cenchrus ciliaris Common	Terán, N.L., México (2001-02)	0.9	1.3	0.6	1.2	1.0
Cenchrus ciliaris Common	Linares, N.L., México (1998-99)	1.1	1.2	1.3	1.1	1.2
Cynodon dactylon II	Marin, N.L., México (1994)	2.0	2.0	2.0	2.0	2.0
Cynodon dactylon	Linares, N.L., México (1998-99)	1.0	1.0	2.0	1.0	1.3
Dichanthium annulatum	Linares, N.L., México (1998-99)	1.0	1.0	0.4	1.0	0.9
Panicum coloratum	Linares, N.L., México (1998-99)	1.0	1.0	1.0	1.0	1.0
Rhynchelytrum repens	Terán, N.L., México (2001-02)	0.8	0.6	0.9	0.8	0.8
Mean		1.1	1.2	1.2	1.2	1.2

Obtained from: Ramirez *et al.* (2002); Ramirez *et al.* (2003); Ramirez-Lozano (2003); Ramirez *et al.* (2004); Ramirez *et al.* (2005).

Table 8.24. Phosphorous content (g kg⁻¹; dry matter) of the hybrid buffelgrass Nueces and five new genotypes of *Cenchrus ciliaris* collected in different dates in Teran county of the state of Nuevo Leon, México

Genotypes	Dates of collection			
	Aug. 1999	Nov. 1999	Nov. 2000	Jun. 2000 (Fertilized)
Cenchrus ciliaris Nueces	0.07	0.10	0.11	0.07
Cenchrus ciliaris 307622	0.07	0.10	0.12	0.03
Cenchrus ciliaris 409252	0.07	0.10	0.10	0.05
Cenchrus ciliaris 409375	0.07	0.10	0.10	0.06
Cenchrus ciliaris 409460	0.08	0.10	0.10	0.05
Cenchrus ciliaris 443	0.06	0.10	0.10	0.11
Mean	0.07	0.10	0.11	0.06

Obtained from: Garcia-Dessommes *et al.* (2003ab)

Cultivated grasses such as *C. ciliaris, C. dactylon, D. annulatum* and *P. coloratum* contained more P in the leaves on the stems. Although the P decreases as the maturity of plant advances, there is great variation between seasons (Figure 8.11). However, differences in water supplies appear to have a higher and more consistent effect that differences in soil temperature on the P content in grasses; since it has been reported that the concentration of P in forage decreases due to drought conditions and increasing aridity at maturity also affect low concentrations of P in forage. Therefore, P is a limiting nutrient in northeastern Mexico and southern Texas, USA for optimal growth and development of grazing ruminants.

Table 8.25. Phosphorous content (g kg-1; dry matter) of 84 new genotypes of *Cenchrus ciliaris* collected in Teran county of the state of Nuevo Leon, Mexico in November 2000.

Genotypes	P, g kg^{-1}	Genotypes	P, g kg^{-1}	Genotypes	P, g kg^{-1}	Genotypes	P, g kg^{-1}
202513	1.0	409185	0.8	409263	1.0	409459	0.8
253261	0.9	409197	0.8	409264	1.0	409460	0.9
307622	1.1	409200	0.9	409266	0.9	409465	0.9
364428	0.7	409219	0.9	409267	1.0	409466	0.9
364439	1.4	409220	0.9	409269	1.0	409472	0.7
364445	0.7	409222	0.9	409270	0.8	409480	0.7
365654	0.8	409223	0.9	409272	0.8	409529	0.7
365702	0.7	409225	0.8	409273	0.7	409691	0.6
365704	1.0	409227	0.9	409275	0.9	409711	1.0
365713	0.7	409228	0.9	409276	0.7	414447	0.7
365728	0.8	409229	0.8	409278	1.1	414451	1.1
365731	0.9	409230	1.0	409279	1.1	414454	1.1
409142	0.8	409232	1.1	409281	1.1	414460	1.1
409151	1.2	409234	1.0	409282	1.2	414467	1.0
409154	0.6	409235	0.9	409284	1.0	414499	0.9
409155	0.6	409238	1.1	409285	1.0	414511	0.9
409157	0.7	409240	0.8	409287	0.8	414512	0.9
409162	0.7	409242	0.9	409288	1.1	414520	1.1
409164	0.9	409252	1.0	409290	1.0	414532	0.9
409165	0.9	409254	0.9	409291	1.0	443	0.9
409168	0.7	409258	1.2	409293	1.0	NUECES	1.1
Mean of all genotypes							0.9

Obtained from: Morales-Rodríguez *et al.*, 2005

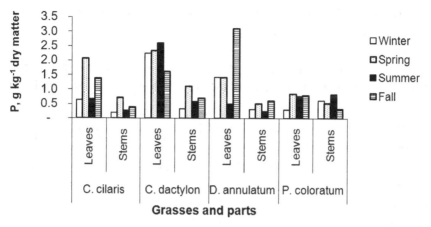

Figure 8.11. Seasonal content of P (g kg-1 dry matter) in leaves
and stems of four cultivated grasses collected in Linares county
of the state of Nuevo Leon, Mexico in 1998 and 1999.
Obtained from: Ramirez *et al.* (2003); Ramirez-Lozano (2003); Ramirez *et al.* (2005a).

Phosphorous content in native grasses

All native grasses listed in Tables 8.26 and 8.27 had P content in lower amounts to satisfy the metabolic requirements of ruminants. Only grasses such as *Cenchrus incertus*, *Panicum obtusum*, *Leptochloa filiformis*, *Setaria macrostachya* and *Panicum unispicatum* contained P in marginally quantities to cover maintenance requirements for all classes of beef cattle. However, P content, in all native grasses (Tables 6.26 and 6.27), during winter was substantially lower than in other seasons.

The P availability to grasses may be limited by its low abundance in the soil, but also, and very commonly, by its adsorption into various soil minerals. Phosphorus deficiency is often the major limitation to crop growth on most soils of northeastern Mexico, particularly where previous cropping has caused a depletion of soil organic matter and increased acidification. Phosphorus deficiency is also common on highly weathered tropical soils and siliceous sands; in fact, few soils are naturally well endowed with this nutrient. Thus, the world is facing an immediate crisis in the decline in the availability of P mineral to provide for the food needs of globe. Poor availability of P in soils and consequent P-deficiency are major constraints to crop production globally.

Table 8.26. Seasonal phosphorous content (g kg-1, dry matter) in native grasses collected in Marin county of the state of Nuevo Leon, Mexico in 1994

Native grasses	Seasons				Annual
	Winter	Spring	Summer	Fall	mean
Aristida longiseta	0.4	0.6	0.4	0.6	0.5
Bouteloua gracilis	0.5	1.0	0.6	1.1	0.8
Cenchrus incertus	1.0	1.5	1.3	1.8	1.4
Hilaria belangeri	0.2	0.9	0.7	0.9	0.7
Panicum hallii	0.9	1.4	1.0	1.3	1.2
Setaria macrostachya	0.4	1.2	0.9	1.3	1.0
Seasonal means	0.6	1.1	0.8	1.2	0.9

Obtained from: Ramirez et al. (2004)

Table 8.27. Seasonal phosphorous content (g kg-1, dry matter) in native grasses collected in Teran county of the state of Nuevo Leon, Mexico in 2001 and 2002

Native grasses	Seasons				Annual
	Winter	Spring	Summer	Fall	mean
Bouteloua curtipendula	0.8	1.3	0.6	0.9	0.9
Bouteloua trifida	0.7	1.0	0.8	0.4	0.7
Brachiaria fasciculata	0.9	1.2	1.5	1.4	1.3
Digitaria insulares	1.1	1.2	1.5	1.4	1.3
Chloris ciliata	1.0	1.4	1.0	1.1	1.1
Leptochloa filiformis	1.2	1.0	1.8	1.8	1.5
Panicum hallii	1.1	0.6	1.2	1.3	1.1
Panicum obtusum	1.4	1.1	1.4	1.7	1.4
Panicum unispicatum	0.9	1.6	1.7	1.3	1.4
Setaria grisebachii	0.9	1.1	1.3	1.7	1.3
Setaria macrostachya	1.5	1.1	1.5	1.7	1.5
Tridens eragrostoides	1.1	1.3	1.0	1.1	1.1
Tridens muticus	0.5	1.3	0.6	0.4	0.7
Seasonal means	1.0	1.2	1.2	1.2	1.2

Obtained from: Cobio-Nagao (2004).

Sulfur

Sulfur is an essential nutrient for plant growth. Although it is considered a secondary nutrient, it is now becoming recognized as the 'fourth macronutrient', along with N, P and K. Oil crops, legumes, forages and some vegetable crops require S in considerable amounts. In many crops, its amount in the plant is similar to P.

Sulfur is essential for many plant functions. Some of them are

1. A structural component of protein and peptides
2. Active in the conversion of inorganic N into protein
3. Sulfur in organic form is present in the vitamins biotin and thiamine
4. A catalyst in chlorophyll production
5. Promotes nodule formation in legumes
6. A structural component of various enzymes
7. Is part of antioxidant molecules like glutathione and thioredoxin
8. A structural component of the compounds that give the characteristic odors and flavors

Since sulfur in plants is associated with the formation of proteins and chlorophyll, its deficiency symptoms look like those of N. Sulfur is moderately mobile within the plant, therefore deficiency symptoms usually start on the younger leaves and progress over time to the older leaves, resulting in plants becoming uniformly chlorotic. While S deficiency symptoms on an individual leaf look like those of N, N deficiency begins in the lowest leaves, not the newest.

Compounds containing sulfur play a variety of essential functions in the animal body. They act as structural entities (collagen), as catalysts (enzymes), as oxygen carriers (hemoglobin), as hormones (insulin), and as vitamins (thiamine and biotin). Sulfur is present in four amino acids: methionine, cystine, cysteine and taurine. The secondary structure of many proteins is determined by the cross linkage or folding due to covalent disulfide bonds between amino acids.

Sulfur is the element that gives many key compounds their unique functional properties. For example, acetate is linked to coenzyme

A by a thioester linkage to form acetyl coenzyme A. This compound is required for the formation of key metabolic intermediates such as citrate, acetoacetate and malonate. The sulfur in thiamine allows it to serve as a molecule that transfers carbonyl groups. Thiamine plays a key role in the formation of pentose sugars, which are required for ribonucleic acids synthesis and photosynthesis. Biotin, another sulfur-containing B vitamin, acts a carrier for carbon dioxide in carboxylation reactions.

Ruminants may respond to inorganic sulfur supplementation, especially if the diet is high in nonprotein nitrogen. It has been showed that ruminal microorganism are capable of synthesizing all organic sulfur containing compounds essential for life from inorganic sulfur. When urea or other nonprotein nitrogen sources are fed, the diet may become deficient in sulfur. In addition, the nitrogen to sulfur ratio in rumen microbial protein averages 14.5:1. The common recommendation for the nitrogen:sulfur ratio is 10:1 in diets containing high levels of urea.

CHAPTER 9

Microminerals in grasses

Introduction

Microminerals are involved in many enzyme systems, which make efficient use of the main metabolic dietary nutrients, e.g. energy substances, fiber and protein. They are a set of elements present in traces amounts in the body, with concentrations ranging from micrograms to picograms per gram of wet tissue, some of which are present by chance and seem to have special functions, others are vital for certain important functions (Cu, Zn, Mn, Fe, Co, Se, I, Table 9.1). Despite found as traces, they are essential for maintaining normal metabolism. The trace minerals are used in the synthesis of vitamins, hormone production, enzymatic activity, collagen formation, tissue synthesis, oxygen transport and many other physiological processes related to growth, health and reproduction. It cannot be establish general rules as to their availability in the grasses since they depend largely on the soil where they are grown, but it has been reported that the animals required in smaller amounts to 1.0 mg kg^{-1} dry matter in the diet.

Table 9.1. Chemical symbol and name of microminerals in alphabetic order

Symbol	Name	Atomic number	Atomic mass, g/mol	Density to 20 °C, g/cm³	Fusion point, °C	Ebullition point, °C
Cu	Cooper	29	63.546	8.92	1083.3	2595
Fe	Iron	26	55.847	7.87	1535	2750
I	Iodine	53	126.90447	4.94	113.5	184.4
Mn	Manganese	25	54.93805	7.4	1244	2097
Mo	Molybdenum	42	95.94	10.28	2617	5560
Se	Selenium	34	78.96	4.32	217	685
Zn	Zinc	30	65.39	7.14	419.6	907

Physiological characteristics

Microminerals are components of many tissues and one or more enzyme activities and their deficiencies lead to a variety of pathological consequences and metabolic defects. A number of elements that are not required (or required only in small amounts) can cause toxicity in beef cattle. The maximum concentration of minerals is defined as one dietary level, given for a limited period, shall not prevent animal behavior and generates no toxic residues in human food derived from animal.

The physiological basis of the deficiency of trace minerals is very complex. Some elements are engaged in a particular enzyme, and many others in the absence of one of these elements affect one or more metabolic processes. A deficiency in the diet does not necessarily lead to clinical disease. Several factors predispose to clinical disease in the animal, among which are: 1) the age that deficiency appears, 2) genotypic differences with respect to requirements, 3) discontinuity in demand due to environmental changes, 4) the challenge arises because of concomitant infections or claims in production, 5) individual variations in response to the deficiency, and 6) volume of functional storages.

A deficiency can be divided into four phases:

1. depletion
2. Deficiency
3. Dysfunction
4. Clinical disease

The depletion term describes the failure in the diet to maintain body condition of the trace mineral, and may continue for weeks or months without clinical effects appear when there are substantial body reserves. Depletion occurs when the net requirements of a particular essential element are greater than the net absorption in the intestine. The body in this state may respond by increasing or decreasing the intestinal absorption endogenous losses. There is a decrease in the element in question in the storage sites as the liver; therefore, the plasma concentrations can remain constant.

If dietary deficiency continues, eventually there is a transition to the state of *depletion* deficiency, which is indicated by biochemical indicators indicating that the homeostatic mechanisms cannot maintain constant levels of minerals needed to constant physiological functions. After several periods, concentrations or activities of the enzymes begin to decline until it reaches the stage of *dysfunction*. There may be an additional delayed period, the subclinical stage, before changes in cell functions as manifest a *clinical disease*.

Some minerals, without being in deficiency, when they are administered daily, produce improved individual production by stimulating effect on digestive or a better balance parameter by excess of another mineral, among other possible factors. Therefore, administration of mineral in production systems should be considered not only from a physiological point of view, but also productive. Mineral blocks provided by free access represent a good practice to avoid possible deficiencies, especially of trace minerals.

Copper

Concentrations of Cu in soil solution range from 0.5 to 135 µg/L, depending on techniques used and on soil types. Copper is necessary for chlorophyll formation in plants and assists with several other plant reactions. Plants supplied with good levels of Cu have stronger cell walls and are more resistant to fungal attack. Although plants do have a requirement for Cu, the main reason Cu is applied is for the benefit of grazing stock. Copper, the essential metal in plants, is a constituent of several "key" enzymes, also plays important functions in physiological processes, such as:

1. Photosynthesis and respiration
2. Carbohydrate and nitrate metabolisms
3. Water permeability
4. Reproduction
5. Disease resistance

Plant Cu deficiency is more common on light textured soils such as sands or sandy loams. Where required, Cu is normally applied with the fertilizer at 1-2 kg/ha every 3-6 years. Inclusion of Cu in the fertilizer will provide a long term supply to pasture and grazing stock. Where Cu deficiency has been diagnosed in stock, more direct supplementation such as Cu drenches are recommended. Copper is slightly mobile in plants as it is strongly bound by nitrogen and proteins. With Cu-deficiency, a plant may grow normal vegetation for a short time, but there will be a failure in the redistribution of Cu from old leaves and roots causing chlorosis and tip necrosis in new leaves, known as Cu-deficiency symptoms.

Copper is mainly absorbed through the gastrointestinal tract. From 20 to 60% of the dietary Cu is absorbed, with the rest being excreted through the feces. Once the metal passes through the basolateral membrane it is transported to the liver bound to serum albumin. The liver is the critical organ for Cu homoeostasis. The Cu is partitioned for excretion through the bile or incorporation into intra- and extracellular proteins. The primary route of excretion is through the bile. The transport of Cu to the peripheral tissues is accomplished through the plasma attached to serum albumin, ceruloplasmin or low-molecular-weight complexes (Figure 9.1).

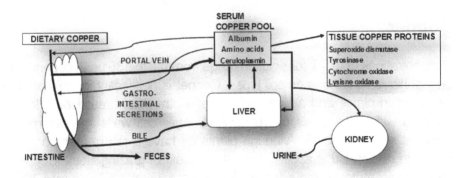

Figure 9.1. Body copper metabolism

The amount of Cu that ruminants absorb from the diet is very variable. The ruminant needs of Cu are remarkably affected by the other components of the diet. Copper needs of beef cattle are normally of 5 mg kg-1 of dry matter of a diet, and significantly increases when there are Mo and S in excess. Excess Cu is stored in the liver. Ruminants are susceptible to Cu toxicity, either if a very large amount of copper is ingested or injected at one time, or if Cu accumulates in the liver over a long period. Copper is required for body, bone and wool growth, for pigmentation, myelination of nerve fibers and leucocyte function.

Throughout the world, Cu deficiency limits cattle production by reducing growth reproductive performance and immune response. With the exception of P, Cu deficiency is the most sever mineral limitation to grazing livestock throughout extensive regions of the tropics. In many geographical areas, Cu deficiency may be caused by an excess of Mo or S that interfere with Cu absorption and function or both. In these areas, Cu consumption may be adequate to meet an animal's needs for Cu, but Mo renders Cu unavailable.

Grazing cattle are more likely to develop a Cu deficiency than feedlot cattle. The extent of Cu released from forages depends upon forage type and stage of growth. Cattle grazing mature low quality forages during the fall and winter are more likely to develop Cu deficiency than those grazing in spring and summer. Two types of soils have a marked influence on Cu uptake by the plants or Cu absorption by the animal. Muck soils, because of their high organic matter bind Cu, which in turn makes forage grown on these soils deficient in Cu. Soils high in Mo can produce forages with high Mo content which can interfere with Cu metabolism in animals grazing these high Mo soils.

Clinical signs of copper deficiency include ill thrift and poor growth in young animals a loss of body condition in the cow and the hair coat may be rough and faded. The change in hair color is the result of loss of pigment in hair follicles. The hair coat will appear more yellow with a red cow and greyer with a black cow. The immune system may also be compromised with animals being more susceptible to infectious diseases and diarrhea. Lameness and in-coordination may be observed in young calves. In older animals, rickets like condition or fractures of long bones may be observed. Degenerative heart disease and associated acute heart failure may also be observed (falling disease). Infertility and delayed or depressed estrus and fetal deaths may also be observed.

Because of the multiple interactions of Cu, Cu deficiency has been grouped into four categories, when the food contains:

1) High levels of Mo (over 20 mg kg^{-1} dry matter)
2) Low level of Cu or Mo considerably high
3) Copper deficiency (less than 5 mg kg-1)
4) Normal level of Cu and low Mo, with a high level of protein soluble (derived from green pastures) since this increases the amount of sulfur produced in the rumen, resulting in Cu sulfide that cannot be use the ruminant.

The Cu, Zn, Fe and Se, along with vitamins C and E act as cofactors in the regulation of free radicals formed in the cells (Figure 7.2). 1) Cu-Zn dismutase reduces the superoxides to H_2O_2, 2) Se-glutathione peroxidase reduces H_2O_2 to unreactive alcohols, 3) Fe catalyzes the production of oxygen radicals such as OH and 4) the vitamins E, C and β- carotenes reduce oxygen radicals to inactive metabolites.

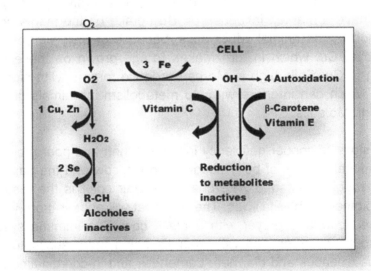

Figure 9.2. Diagram of interaction of the Cu, Fe, Se and the vitamins
C and E to avoid the formation of free radicals in animal cells

Copper content in cultivated grasses

The concentration of Cu present in 94 samples of tropical grasses grown in various parts of the world ranged from 3 to 100 mg kg^{-1} dry matter, with an average of 15 mg kg^{-1} dry matter. The 26% of grasses were deficient in Cu to meet the requirements of growing beef cattle (10 mg kg^{-1} dry matter; Table 9.2). However, this value may be too high, since no variation was found in growth of grazing animals when they consumed Cu at lower levels (3-8 mg kg^{-1} in the dry matter) than the recommended level of 10 mg kg^{-1} in the dry matter.

Table 9.2. Micromineral requirements of ruminants

Element	Beef cattle		Adult sheep	Adult goats
	Growing and finishing	Early lactation		
Cu, mg kg^{-1}	10	10	7 - 11	8 – 10
Fe, mg kg^{-1}	50	50	30 - 50	30 – 40
Mn, mg kg^{-1}	20	40	20 - 40	30 – 40
Zn, mg kg^{-1}	30	30	20 - 33	40 – 50

Obtained from: McDowell (2003).

In many regions of the world, after the P, Cu deficiency is the most important for grazing animals. This is corroborated in cultivated grasses collected in different counties of northeastern Mexico containing Cu levels (Tables 9.3, 9.4, 9.5) insufficient to cover the requirements for growing beef cattle (Table 9.2). In addition, the native grasses that grow in these regions (Tables 9.6 and 9.7), also resulted of Cu deficiency.

Table 9.3. Copper content (mg kg^{-1}; dry matter) in cultivated grasses collected in different counties of the state of Nuevo Leon, México

Grasses	Place and dates of collection	Seasons				Annual
		Winter	Spring	Summer	Fall	mean
Cenchrus ciliaris Common	Teran, N.L., Mexico (2001-02)	5	8	8	9	8
Cenchrus ciliaris Common	Linares, N.L., Mexico (1998-99)	2	2	4	5	3
Cynodon dactylon II	Marin, N.L., Mexico (1994)	4	2	7	5	5
Cynodon dactylon	Linares, N.L., Mexico (1998-99)	9	7	9	9	9
Dichanthium annulatum	Linares, N.L., Mexico (1998-99)	6	4	7	5	6
Panicum coloratum	Linares, N.L., Mexico (1998-99)	5	4	7	4	5
Rhynchelytrum repens	Teran, N.L., Mexico (2001-02)	3	4	6	4	4
Mean		5	5	7	6	6

Obtained from: Ramírez et al. (2002); Ramírez et al. (2003); Ramírez-Lozano (2003); Ramírez et al. (2004); Ramírez et al. (2005).

Table 9.4. Copper content (mg kg-1; dry matter) in the hybrid buffelgrass Nueces and five new genotypes of Cenchrus ciliaris collected in different dates in Teran county of the state of Nuevo Leon, México

Genotypes	Fechas de colecta			
	Aug. 1999	Nov. 1999	Nov. 2000	Jun. 2000 (Fertilized)
Cenchrus ciliaris Nueces	1	3	3	2
Cenchrus ciliaris 307622	2	2	3	2
Cenchrus ciliaris 409252	1	1	3	2
Cenchrus ciliaris 409375	1	1	2	3
Cenchrus ciliaris 409460	1	1	4	2
Cenchrus ciliaris 443	1	2	3	3
Mean	1	2	3	2

Obtained from: García-Dessommes et al. (2003ab)

It appears that urea fertilizer, which has little effect on soil pH, generally has little effect on the concentration of Cu in the forage. This fact has been confirmed in six new genotypes of Cenchrus ciliaris (Table 9.4) with and without fertilization with 120 kg ha^{-1} of urea N, which had no substantial differences in the concentration of Cu. The 84 new genotypes of Cenchrus ciliaris listed in Table 9.4 had a Cu composition very similar to the genotypes on Table 7.4. Moreover, all grasses had low Cu content to meet the metabolic demands of growing ruminant (10 mg kg^{-1} in the dry matter of diet).

Except for grass *C. dactylon*, all grasses that are shown in Figure 9.3, leaves contained more Cu than stems. Maturity in pastures promotes decreased Cu content in cultivated grasses and reduces the difference between plant parts. It has been reported that there is a rapid consumption during growth and a gradual dilution as the plant matures. For example, Cu levels in the forage can be reduced by 50 % as plants mature.

Table 9.5. Copper content (mg kg-1; dry matter) of 84 new genotypes of *Cenchrus ciliaris* collected in Teran county of the state of Nuevo Leon, Mexico in November 2000

Genotypes	Cu, mg kg⁻¹	Genotypes	Cu, mg kg⁻¹	Genotypes	Cu, mg kg⁻¹	Genotypes	Cu, mg kg⁻¹
202513	4	409185	3	409263	3	409459	3
253261	4	409197	3	409264	1	409460	3
307622	3	409200	2	409266	1	409465	3
364428	3	409219	3	409267	2	409466	4
364439	2	409220	3	409269	2	409472	2
364445	3	409222	3	409270	2	409480	2
365654	4	409223	5	409272	2	409529	2
365702	3	409225	2	409273	1	409691	2
365704	4	409227	3	409275	3	409711	2
365713	2	409228	3	409276	1	414447	2
365728	4	409229	3	409278	6	414451	2
365731	5	409230	3	409279	2	414454	2
409142	5	409232	3	409281	2	414460	4
409151	2	409234	2	409282	2	414467	2
409154	2	409235	2	409284	2	414499	2
409155	3	409238	4	409285	3	414511	3
409157	3	409240	3	409287	1	414512	2
409162	3	409242	2	409288	2	414520	2
409164	2	409252	2	409290	2	414532	3
409165	3	409254	3	409291	2	443	1
409168	3	409258	3	409293	3	Nueces	4
Mean of all genotypes							3

Obtained from: Morales-Rodríguez *et al.*, 2005

Copper content in native grasses

Of all the native grasses listed in Table 9.6 and Table 9.7, *Panicum hallii*, collected in Marin county of the state of Nuevo Leon, Mexico, was the only grass, in all seasons of the year (Table 7.6), that had levels Cu to meet the requirements of growing beef cattle (Table 9.2). Differences in the concentration of Cu may be due to:

1. Differences between plant species
2. Maturity of the plant
3. Parts of the plant
4. Type of soil and mineral contained within soil
5. Seasonal variations

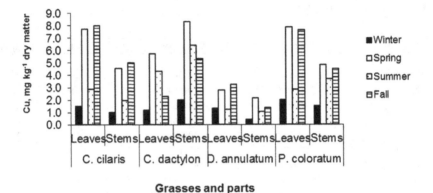

Figure 9.3. Seasonal content of Cu (mg kg-1 dry matter) in leaves and stems of four cultivated grasses, collected in Linares county of the state of Nuevo Leon, Mexico in 1998 and 1999.

Obtained from: Ramírez *et al.* (2003); Ramírez-Lozano (2003); Ramírez *et al.* (2005a).

Table 9.6. Seasonal content of Cu (mg kg-1 dry matter) in native grasses collected in Marin county of the state of Nuevo Leon, Mexico in 1994

| Native grasses | Seasons | | | | Annual |
	Winter	Spring	Summer	Fall	mean
Aristida longiseta	4	6	6	9	6
Bouteloua gracilis	3	8	4	10	6
Cenchrus incertus	7	9	7	12	9
Hilaria belangeri	4	7	5	10	7
Panicum hallii	11	12	11	11	11
Setaria macrostachya	7	8	5	9	7
Seasonal means	6	8	6	10	7

Obtained from: Ramirez *et al.* (2004).

Table 9.7. Seasonal content of Cu (mg kg-1 dry matter) in native grasses collected in Teran county of the state of Nuevo Leon, Mexico in 2001 and 2002

| Native grasses | Seasons | | | | Annual |
	Winter	Spring	Summer	Fall	mean
Bouteloua curtipendula	2	2	1	2	2
Bouteloua trifida	3	2	2	2	2
Brachiaria fasciculata	3	4	8	2	4
Digitaria insularis	3	3	6	4	4
Chloris ciliata	2	6	3	2	3
Leptochloa filiformis	6	5	2	3	4
Panicum hallii	3	2	6	3	3
Panicum obtusum	6	5	3	4	5
Panicum unispicatum	4	6	4	4	4
Setaria grisebachii	2	3	7	3	4
Setaria macrostachya	2	3	8	3	4
Tridens eragrostoides	4	5	3	3	4
Tridens muticus	2	2	2	2	2
Seasonal means	3	4	4	3	3

Obtained from: Cobio-Nagao (2004).

Iron (Fe)

Iron is a necessary trace element found in nearly all living organisms. Iron-containing enzymes and proteins, often containing heme prosthetic groups, participate in many biological oxidations and in transport. Iron is involved in the production of chlorophyll, and Fe chlorosis is easily recognized on Fe-sensitive crops growing on calcareous soils. Iron also is a component of many enzymes associated with energy transfer, nitrogen reduction and fixation, and lignin formation. Iron is associated with sulfur in plants to form compounds that catalyze other reactions. The essential role of Fe in plant biochemistry may be summarized as follows:

1. Several Fe-proteins, mainly transferrins, ferritins, and siderophores, are involved in transport, storage, and binder systems.

2. Fe occurs in hem and nonhem proteins and is concentrated mainly in chloroplasts.
3. Fe influences chlorophyll formation.
4. Organic Fe complexes are involved in the mechanisms of photosynthesis electron transfer.
5. Nonhem Fe proteins are involved in the reduction of nitrites and sulfates.
6. Fe is directly implicated in the metabolism of nucleic acids.
7. Both cations, Fe^{3+} and Fe^2, may also play a catalytic role in various reactions.

Yellow leaves due to low levels of chlorophyll mainly manifest iron deficiencies. Leaf yellowing first appears on the younger upper leaves in interveinal tissues. Severe Fe deficiencies cause leaves to turn completely yellow or almost white, and then brown as leaves die. Iron deficiencies are found mainly on high pH soils, although some acid, sandy soils low in organic matter also may be iron-deficient. Cool, wet weather enhances iron deficiencies, especially on soils with marginal levels of available iron. Poorly aerated or compacted soils also reduce iron uptake by plants. Uptake of iron decreases with increased soil pH, and is adversely affected by high levels of available phosphorus, manganese and zinc in soils.

In animals, Fe is important for red blood cell function because of its fundamental role in the hemoglobin molecule. Hemoglobin is the molecule that carries oxygen to the tissues from the lungs and returns with carbon dioxide from the cells. It is also closely related to myoglobin, which is responsible for energy production in the muscles.

Main functions of Fe in the body are:

1. Main component of hemoglobin, which transports oxygen and carbon dioxide to and from the lungs.
2. Supports the action of many enzymes (especially for energy production).
3. Antioxidant.
4. May have anti-cancer properties.
5. Powerful immune-system booster.

Although iron is the second most abundant mineral on our planet and is present in many food sources, iron deficiency is still the most common mineral deficiency. Often this is related to the many factors which either increase or decrease iron absorption, as well as the absolute amount of iron in the diet. As suggested above, the form in which iron is taken has a significant effect on its bioavailability, and the hem-form is much better absorbed and utilized by the body.

Many of the symptoms of iron deficiency are related to the consequences of poor oxygen and carbon dioxide transport. For example, children who are iron deficient may have learning difficulties due to the fact that their brains are "starved" (relatively speaking) of oxygen, which makes the brain less efficient. Anemia may lead to lack of exercise which, in turn affects fitness and other body systems.

In the ruminant over 90% of the Fe existing in the body is combined with proteins, especially hemoglobin. It is also found in blood plasma joined the transferritin protein, which carries the Fe in the body. It is stored as ferritin in liver, spleen, kidney and bone marrow, or as hemosiderin. Also, form part of many enzymes, including cytochromes and flavoproteins (Figure 9.4).

Iron is absorbed in the intestinal lumen by the mucosal cells. The absorption is related to the functional needs and is more efficient in young animals than in adults. The heme compounds in foods of animal origin such as fish meal, are better absorbed than iron from plant foods, which contain mainly inorganic salts. The extent of absorption is affected by the chelated, some of which (ascorbic acid or cysteine) favor absorption, while others inhibit. Iron absorption is reduced by other divalent ions (Zn, Mn, Co) considered compete for the binding sites on the intestinal mucosa. Phosphates and phytates interfere with the absorption of iron to form insoluble iron salts. Copper positively interferes in iron utilization, since Cu is part of the ferroxidase enzyme that facilitates the release of Fe from ferritin in the cells of the intestinal mucosa.

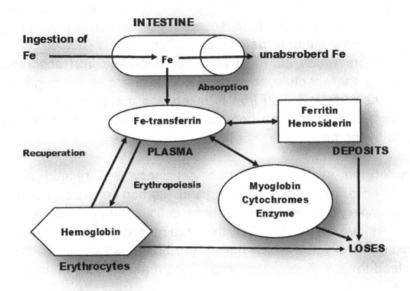

Figure 9.4. Ingestion and transport, deposit and losses of Fe in animal organism

The requirements of Fe (25-40 mg kg^{-1}; dry basis) are low in adult ruminant diets. The administration of compounds of Fe to pregnant females may serve to increase levels of blood hemoglobin and Fe stores of newborn animals, but does not increase the Fe content of milk. Anemia is the main symptom of iron deficiency depleted its reserves in the body characterized by the reduction content of erythrocytes and hemoglobin in blood.

Iron content in cultivated grasses

Young animals require more Fe than adults. Fe deficiencies for grazing ruminants are rare unless bleeding occurs (by parasites or disease). Apparently most cultivated grasses listed in Tables 9.8, 9.9 and 9.10 contain concentrations of Fe in sufficient quantities to meet the needs of growing cattle (50 mg kg^{-1} dry matter of the diet; Table 9.2). However, there is variation in the seasonal trend in the Fe content in cultivated grasses; in winter and spring (dry seasons), grasses had lower Fe content than in summer and fall (wet seasons). This fact may be explained due to there is evidence that the ability of soil to support a certain level of productivity in a system, is not constant and tends to fluctuate over time. Such fluctuation affects the physical composition, chemical and soil fertility as the availability of minerals therein. This situation is reflected natural form when there are variations in the production of green forage, alterations of the normal cycles of growth of grasses and low capacity to adapt themselves to adverse environmental conditions.

Table 9.8. Seasonal Fe content (mg kg^{-1} dry matter) in cultivated grasses collected in different dates and counties of the state of Nuevo Leon, México

Cultivated grasses	Place and dates of collection	Seasons				Annual
		Winter	Spring	Summer	Fall	mean
Cenchrus ciliaris Common	Teran, N.L., Mexico (2001-02)	114	164	120	176	144
Cenchrus ciliaris Common	Linares, N.L., Mexico (1998-99)	91	164	163	255	168
Cynodon dactylon II	Marin, N.L., Mexico (1994)	120	81	183	129	128
Cynodon dactylon	Linares, N.L., México (1998-99)	73	66	85	78	76
Dichanthium annulatum	Linares, N.L., Mexico (1998-99)	96	88	132	105	105
Panicum coloratum	Linares, N.L., Mexico (1998-99)	113	76	127	92	102
Rhynchelytrum repens	Teran, N.L., Mexico (2001-02)	53	77	135	97	91
Seasonal means		94	102	135	133	116

Obtained from: Ramirez et al. (2002); Ramirez et al. (2003); Ramirez-Lozano (2003); Ramirez et al. (2004); Ramirez et al. (2005).

Table 9.9. Seasonal Fe content (mg kg-1 dry matter) in the hybrid buffelgrass Nueces and five new genotypes of Cenchrus ciliaris collected in different dates in Teran county of the state of Nuevo Leon, Mexico

Genotypes	Dates of collection			
	Aug. 1999	Nov. 1999	Nov. 2000	Jun. 2000 (Fertilized)
Cenchrus ciliaris Nueces	115	119	138	126
Cenchrus ciliaris 307622	125	131	172	148
Cenchrus ciliaris 409252	73	90	76	94
Cenchrus ciliaris 409375	108	116	135	215
Cenchrus ciliaris 409460	112	119	132	126
Cenchrus ciliaris 443	89	104	115	152
Mean	104	110	128	144

Obtained from: Garcia-Dessommes et al. (2003ab).

In grazing systems, natural suppliers of minerals for livestock are grasses and drinking water. Grasses, in turn, are extracted from removable compounds in the soil where they grow, this why their presence and availability are critical in farms productive system based on grazing. It seems that, in alkaline soils, such as those from Teran county of the state of Nuevo Leon, Mexico, nitrogen fertilization in the form of urea had no effect on the Fe content in new genotypes of buffelgrass. It has been reported that by applying 120 kg ha^{-1} of urea on the soil, the Fe content in grasses increased slightly compared with unfertilized grasses (Table 9.10).

Table 9.10. Iron content (mg kg^{-1}; dry matter) in 84 new genotypes of *Cenchrus ciliaris* collected in Teran county of the state of Nuevo Leon, Mexico in November 2000

Genotypes	Fe, mg kg^{-1}	Genotypes	Fe, mg kg^{-1}	Genotypes	Fe, mg kg^{-1}	Genotypes	Fe, mg kg^{-1}
202513	172	409185	98	409263	111	409459	198
253261	164	409197	94	409264	104	409460	195
307622	158	409200	133	409266	100	409465	130
364428	179	409219	145	409267	187	409466	177
364439	93	409220	127	409269	102	409472	171
364445	84	409222	132	409270	180	409480	119
365654	97	409223	151	409272	151	409529	111
365702	157	409225	144	409273	83	409691	119
365704	186	409227	126	409275	128	409711	92
365713	151	409228	129	409276	112	414447	169
365728	114	409229	119	409278	77	414451	184
365731	154	409230	152	409279	81	414454	199
409142	157	409232	136	409281	119	414460	126
409151	95	409234	148	409282	178	414467	164
409154	128	409235	140	409284	111	414499	124
409155	94	409238	107	409285	158	414511	133
409157	149	409240	113	409287	138	414512	132
409162	115	409242	137	409288	188	414520	143
409164	134	409252	130	409290	198	414532	111
409165	110	409254	122	409291	113	443	157
409168	76	409258	92	409293	130	Nueces	101
Mean of all genotypes							132

Obtained from: Morales-Rodriguez *et al.*, 2005b

Even though, the 84 new genotypes of *Cenchrus ciliaris,* listed in Table 9.10, contained variable concentrations of Fe, all had sufficient quantities to meet the demands of growing beef cattle. In addition, grazing animals receive additional amounts of Fe by ingestion of contaminated soil particles that are rich in Fe.

Fe concentrations in animals and plants are very similar, approximately from 100 to 150 mg kg⁻¹ dry basis. However, that the Fe requirements for growing ruminants is much greater than those in maintenance. It appears that, the Fe content in leaves is twice the stems on grasses *C. ciliaris*, *C. dactylon*, *D. annulatum* and *P. coloratum* (Figure 9.5), with little variation among grasses and between seasons. Therefore, leaves of the above grasses represent a good source of Fe for adult ruminants.

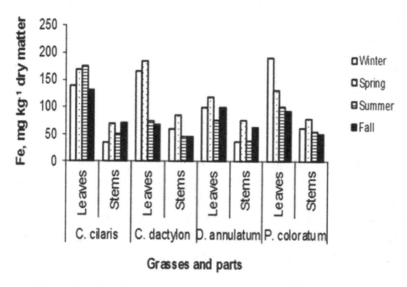

Figure 9.5. Seasonal Fe content (mg kg-1 dry matter) in leaves and stems in four cultivated grasses collected in Linares county of the state of Nuevo Leon, Mexico in 1998 and 1999.
Obtained from: Ramirez *et al.* (2003); Ramirez-Lozano (2003); Ramirez *et al.* (2005a).

Iron content in native grasses

Even though, Fe content varied among season and between species in all native grasses growing in Marin county of the state of Nuevo Leon, Mexico (Table 9.11), all had concentrations to meet the metabolic demands of adult ruminants (50 mg kg⁻¹ dry matter of the diet). *Aristida longiseta* had the highest content of Fe and *Hilaria belangeri* was the lowest. During wet seasons (summer and fall) all grasses had more Fe than in dry seasons (winter and spring).

Table 9.11. Seasonal Fe content (mg kg-1 dry matter) in native grasses collected in Marin county of the state of Nuevo Leon, Mexico in 1994

Native grasses	Seasons				Annual
	Winter	Spring	Summer	Fall	means
Aristida longiseta	190	193	141	192	179
Bouteloua gracilis	100	111	101	119	108
Cenchrus incertus	72	181	94	188	134
Hilaria belangeri	79	89	85	124	94
Panicum hallii	109	166	117	187	145
Setaria macrostachya	83	138	117	102	110
Seasonal means	106	146	109	152	128

Obtained from: Ramirez *et al.* (2004)

Native grasses that grow in the Teran county of the state of Nuevo Leon, Mexico also had Fe concentrations that varied between plant species and among seasons (Table 9.12). However, all grasses had, in all seasons, concentrations of Fe to meet the demands of adult ruminants (50 mg kg^{-1} dry matter of the diet). *Bouteloua curtipendula* had the lowest Fe content and *Panicum obtusum* was highest.

Table 9.12. Seasonal Fe content (mg kg-1 dry matter) in native grasses collected in Teran county of the state of Nuevo Leon, Mexico in 2001 and 2002

Native grasses	Seasons				Annual
	Winter	Spring	Summer	Fall	means
Bouteloua curtipendula	56	66	92	84	74
Bouteloua trifida	79	78	189	75	105
Brachiaria fasciculata	100	128	153	175	139
Digitaria insularis	119	198	150	165	158
Chloris ciliata	58	108	143	191	125
Leptochloa filiformis	122	122	122	122	122
Panicum hallii	188	126	188	180	170
Panicum obtusum	188	176	188	181	183
Panicum unispicatum	116	135	152	185	147

Setaria grisebachii	104	90	147	122	116
Setaria macrostachya	91	75	155	99	105
Tridens eragrostoides	148	108	173	110	135
Tridens muticus	117	96	137	113	115
Seasonal means	114	116	153	138	130

Obtained from: Cobio-Nagao (2004)

Manganese

All plants have a specific requirement for Mn and apparently the most important Mn function is related to the oxidation–reduction processes. The Mn in plants is known to be a specific component of two enzymes, arginase and phosphotransferase, but this metal can also substitute for Mg in other enzymes. The mechanism by which Mn^{2+} activates several oxidases is not yet known precisely, but it appears to be related to the valency change between Mn^{3+} and Mn^{2+}.

In addition, Mn appears to participate in the O_2-evolving system of photosynthesis and also plays a basic role in the photosynthetic electron transport system. Apparently, the Mn fraction that is loosely bound in chloroplasts is associated with O_2 evolution, whereas the firmly bound Mn fraction is involved in the electron pathway in photosynthesis. It was described Mn as a key to life and emphasized its role in O_2 evolution in the photosynthesis. The role of Mn in the NO_2–reduction step is not yet clear, but it appears to be a kind of indirect relationship between the Mn activity and N assimilation by plants. Chloroplasts are the most sensitive of all cell components to Mn deficiency and react by showing structural impairment. Deficiency symptoms occur first in younger leaves as interveinal chlorosis.

Manganese is an important metal for animal health, being necessary for development, metabolism, and the antioxidant system (Figure 9.6). The classes of enzymes that have manganese cofactors are very broad, and include oxidoreductases, transferases, hydrolases, lyases, isomerases, ligases, lectins, and integrins. The consequences of Mn deficiency are skeletal malformations, growth retardation, reproductive disorders and abnormalities in newborns. Supplementation of Fe and Mn is less important in tropical regions where most soils are acidic.

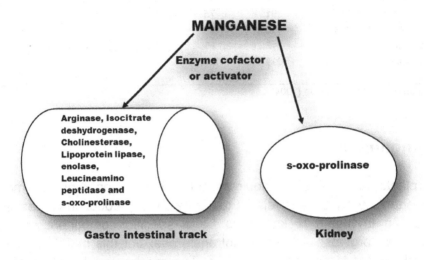

Figure 9.6. Schematic role of Manganese as enzyme activator or cofactor

Manganese content in cultivated grasses

All cultivated grasses harvested in Teran county of the state of Nuevo Leon, Mexico contained sufficient Mn, in all seasons, to meet the metabolic needs of growing beef cattle (20 mg kg^{-1} dry matter of the diet); although *C. ciliaris and R. repens* (Table 9.13) were marginally sufficient. In summer when rainfall in the region was more abundant, grasses had higher Mn content. During the other seasons, the Mn was lower and similar between them.

It appears that the Mn content in the grasses that grow in acidic soils is not affected by N fertilization in the form of urea. However, six new genotypes of *Cenchrus ciliaris* had more Mn content when were fertilized than the same unfertilized genotypes (Table 9.13). This fact might have been due the soils in Teran county of the state of Nuevo Leon, Mexico are slightly alkaline. In addition, all genotypes in all collection dates have adequate Mn for the requirements of growing ruminants.

Table 9.13. Seasonal Mn content (mg kg⁻¹ dry matter) in cultivated grasses collected in different dates in different counties of the state of Nuevo Leon, Mexico

Cultivated grasses	Place and dates of collection	Seasons				Annual
		Winter	Spring	Summer	Fall	mean
Cenchrus ciliaris Common	Terán, N.L., México (2001-02)	27	44	33	39	36
Cenchrus ciliaris Common	Linares, N.L., México (1998-99)	20	22	22	19	21
Cynodon dactylon II	Marin, N.L., México (1994)	32	28	42	37	35
Cynodon dactylon	Linares, N.L., México (1998-99)	54	36	58	60	55
Dichanthium annulatum	Linares, N.L., México (1998-99)	52	34	62	67	54
Panicum coloratum	Linares, N.L., México (1998-99)	33	25	54	26	35
Rhynchelytrum repens	Terán, N.L., México (2001-02)	30	24	29	27	28
Seasonal means		35	30	43	39	37

Obtained from: Ramirez et al. (2002); Ramirez et al. (2003); Ramirez-Lozano (2003); Ramirez et al. (2004); Ramirez et al. (2005).

Table 9.14. Seasonal Mn content (mg kg-1 dry matter) in the hybrid buffelgrass Nueces and five new genotypes of Cenchrus ciliaris collected in different dates in Teran county of the state of Nuevo Leon, Mexico

Genotypes	Collection dates			
	Aug. 1999	Nov. 1999	Nov. 2000	Jun. 2000 (Fertilized)
Cenchrus ciliaris Nueces	25	31	29	38
Cenchrus ciliaris 307622	24	32	30	42
Cenchrus ciliaris 409252	28	39	30	42
Cenchrus ciliaris 409375	32	38	30	50
Cenchrus ciliaris 409460	32	34	29	50
Cenchrus ciliaris 443	16	18	42	49
Means	26	32	32	40

Obtained from: Garcia-Dessommes et al. (2003ab).

It has been reported that concentrations of Mn in forages worldwide range from 1.0 to 2670 mg kg⁻¹ dry matter, with a value of 86 mg kg⁻¹. The 84 genotypes shown in Table 9.15 that were planted under rainfed conditions, without fertilization, had Mn concentrations that were lower than the world mean; however, they were sufficient for the metabolic needs of ruminants.

It appears that there is not consistent pattern in the Mn content between leaves and stems of grasses. It also has been mentioned that the variability between parts of the plant might be attributed to the Mn availability in soils. However, grasses such as C. *ciliaris, C. dactylon, D. annulatum* and *P. coloratum* Mn had higher Mn content in the leaves than in the stems (Figure 9.7).

Manganese content in native grasses

Even though the Mn content was variable between species of native grasses that grow in the northeast of Mexico, there is no clear seasonal trend in the Mn concentration (Tables 9.16 and 9.17). This fact was also previously reported in several previous studies. It seems that, all native grasses, in all seasons, had enough Mn to meet the needs of growing ruminants (Table 9.2). Forages generally contain adequate manganese, assuming the Mn is available for absorption. Studies on range grasses at Vieh ranch in US, have reported that Mn content was always at a sufficient level to potentially provide the beef cow and sheep ewe with an adequate amount of Mn to meet their needs.

Table 9.15. Manganese content (g kg^{-1}; dry matter) of 84 of new genotypes of *Cenchrus ciliaris* collected in Teran county of the state of Nuevo Leon, Mexico in November 2000

Genotypes	Mn, g kg^{-1}	Genotypes	Mn, g kg^{-1}	Genotypes	Mn, g kg^{-1}	Genotypes	Mn, g kg^{-1}
202513	42	409185	25	409263	49	409459	43
253261	34	409197	40	409264	43	409460	31
307622	44	409200	40	409266	37	409465	37
364428	35	409219	36	409267	37	409466	42
364439	29	409220	34	409269	35	409472	27
364445	24	409222	47	409270	38	409480	21
365654	37	409223	37	409272	37	409529	20
365702	24	409225	36	409273	28	409691	23
365704	33	409227	33	409275	33	409711	22
365713	29	409228	36	409276	31	414447	26
365728	37	409229	38	409278	25	414451	26
365731	39	409230	43	409279	34	414454	26
409142	33	409232	39	409281	68	414460	39
409151	29	409234	44	409282	38	414467	30
409154	24	409235	40	409284	40	414499	30
409155	30	409238	41	409285	38	414511	30
409157	36	409240	40	409287	27	414512	34
409162	28	409242	41	409288	30	414520	29
409164	29	409252	63	409290	38	414532	35
409165	31	409254	36	409291	26	443	74
409168	26	409258	48	409293	40	NUECES	42
Mean of all genotypes							35

Obtained from: Morales-Rodríguez *et al.*, 2005b

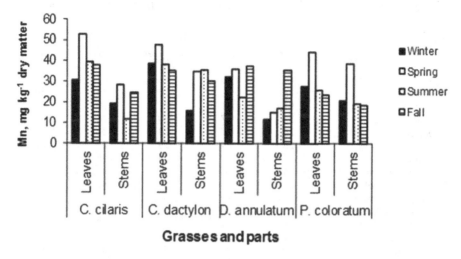

Figure 9.7. Seasonal Mn content (mg kg-1 dry matter) in
leaves and stems of four cultivated grasses
Collected in Linares county of the state of Nuevo Leon, Mexico in 1998 and 1999.
Obtained from: Ramirez *et al.* (2003); Ramirez-Lozano (2003); Ramirez *et al.* (2005a).

Table 9.16. Seasonal Mn content (mg kg-1 dry matter) in native grasses
collected in Marin county of the state of Nuevo Leon, Mexico in 1994

| Native grasses | Seasons | | | | Annual |
	Winter	Spring	Summer	Fall	mean
Aristida longiseta	31	44	31	49	39
Bouteloua gracilis	28	48	41	60	39
Cenchrus incertus	50	70	65	74	65
Hilaria belangeri	34	38	33	44	37
Panicum hallii	38	57	55	71	55
Setaria macrostachya	33	43	39	48	41
Seasonal means	36	50	44	58	46

Obtained from: Ramirez *et al.* (2004)

Table 9.17. Seasonal Mn content (mg kg-1 dry matter) in native grasses collected in Teran county of the state of Nuevo Leon, Mexico in 2001 and 2002

| Native grasses | Seasons | | | | Annual mean |
	Winter	Spring	Summer	Fall	
Bouteloua curtipendula	46	40	49	47	45
Bouteloua trifida	38	30	38	36	36
Brachiaria fasciculata	54	31	78	89	63
Digitaria insulares	70	67	80	76	88
Chloris ciliata	28	27	23	33	27
Leptochloa filiformis	25	24	29	22	25
Panicum hallii	46	38	45	45	43
Panicum obtusum	34	32	34	38	35
Panicum unispicatum	25	46	48	42	40
Setaria grisebachii	31	29	35	30	31
Setaria macrostachya	31	36	40	33	35
Tridens eragrostoides	38	31	44	40	38
Tridens muticus	22	29	30	28	27
Seasonal means	38	35	44	43	41

Obtained from: Cobio-Nagao (2004)

Zinc

Zinc plays essential metabolic roles in grasses, of which the most significant is its activity as a component of a variety of enzymes, such as dehydrogenases, proteinases, peptidases, and phosphohydrolases. It has been showed that the elementary Zn purposes in plants are related to the metabolism of carbohydrates, proteins, and phosphates, and to auxins, RNA, and ribosome formations. There are evidences that Zn influences the permeability of membranes and that stabilizes cellular components and systems of microbes. It appears that Zn stimulate the resistance of plants to dry and hot weather and to bacterial and fungal diseases. This may be because the positive Zn effects on levels of proteins and chlorophyll and abscisic acid.

Plant species and varieties differ extensively in their vulnerability to Zn deficiencies. Although these deficiencies are relatively common, their analysis is rather complex and the best diagnoses are obtained when based on visual symptoms, plant analyses, and soil testing together. It must be emphasized, however, that grasses and soils, two chelating extractants, DTPA and EDTA, give linear relationships between Zn in plants and the soluble Zn pool the soils.

Zinc is widely distributed throughout the body as a component of metalloenzymes and metalloproteins. Zinc finger proteins play an integral role in regulating gene expression, consequently influencing a wide variety of body functions including cell division, growth, hormone production, metabolism, appetite control, and immune function. Zinc has a catalytic, coactive, or structural role in a wide variety of enzymes that regulate many physiological processes including metabolism, growth, and immune function. Zinc play an important role as a cofactor of the enzyme superoxide dismutase. Because Zn is required for production of protective keratins in the hoof and teat, one area of recent attention has been evaluating the role Zn plays in maintaining structural integrity and health of the hoof and udder. The recommended requirement of Zn in beef cattle diets is 30 mg Zn/kg diet, a concentration that should satisfy requirements in most situations.

Zinc, Cu, Mn, and Se are important to a variety of biological processes. Organic forms of trace mineral supplements are generally more bioavailable than inorganic forms, although bioavailability measures vary substantially among trials. There is some evidence that organic forms of trace minerals improve production and health responses of dairy cattle relative to inorganic trace minerals. Supplementation of Zn, Cu, Mn, or Se during times of oxidative stress may reduce oxidative damage to white blood cells and increase disease resistance.

Zinc content in cultivated grasses

The concentration of Zn in 119 samples of tropical grasses grown in different parts of the world ranged from 15 to 120 mg Zn kg^{-1}, with a mean of 36 mg Zn kg^{-1}. The differences between grass species may be due to differences in the content of Zn in soils, differences between species and stage of maturity of grasses. However, generally warm climate grasses tend to have lesser Zn than template climate grasses, and grasses are generally lower than legumes.

The level of Zn in the diet of growing beef cattle is 30 mg kg^{-1} dry matter for not suffering deficiency symptoms. All cultivated grasses that are shown in Table 9.18, had sufficient amounts of Zn to meet the metabolic requirements, in all seasons, of growing beef cattle. Although there were differences between grass species, there were no clear differences between stations in the concentration of Zn. This fact has been previously reported.

Table 9.18. Seasonal Zn content (mg kg^{-1} dry matter) in cultivated grasses collected in different dates of some counties of the state of Nuevo Leon, Mexico

| Cultivated grasses | Place and dates of collection | Seasons | | | | Annual |
		Winter	Spring	Summer	Fall	mean
Cenchrus ciliaris Common	Teran, N.L., Mexico (2001-02)	41	52	43	71	52
Cenchrus ciliaris Common	Linares, N.L., Mexico (1998-99)	34	38	70	61	51
Cynodon dactylon II	Marin, N.L., Mexico (1994)	37	31	45	33	37
Cynodon dactylon	Linares, N.L., Mexico (1998-99)	40	39	63	50	48
Dichanthium annulatum	Linares, N.L., Mexico (1998-99)	40	39	44	41	41
Panicum coloratum	Linares, N.L., Mexico (1998-99)	37	30	42	28	34
Rhynchelytrum repens	Teran, N.L., Mexico (2001-02)	50	51	39	43	46
Seasonal means		40	40	49	47	44

Obtained from: Ramirez *et al.* (2002); Ramirez *et al.* (2003); Ramirez-Lozano (2003); Ramirez *et al.* (2004); Ramirez *et al.* (2005).

The influence of N fertilization on the concentration of Zn in forages, depends mainly if the fertilizer changes the pH of the soil. Apparently, fertilization with N as urea had no influence on the pH of the soil of Teran county of the state of Nuevo Leon, Mexico due to the concentration of Zn was similar between fertilized (120 kg urea ha^{-1}) and unfertilized genotypes (Table 9.19). Moreover, they had insufficient Zn content for the requirements of growing beef cattle (30 mg kg-1 dry matter of the diet).

Table 9.19. Seasonal Zn content (mg kg-1 dry matter) in the hybrid buffelgrass Nueces and five new genotypes of *Cenchrus ciliaris* collected in different dates of Teran county of the state of Nuevo Leon, México

| Genotypes | Dates of collection | | | |
	Aug. 1999	Nov. 1999	Nov. 2000	Jun. 2000 (Fertilized)
Cenchrus ciliaris Nueces	13	12	12	13
Cenchrus ciliaris 307622	11	12	19	12
Cenchrus ciliaris 409252	14	17	26	15
Cenchrus ciliaris 409375	10	17	28	15

Cenchrus ciliaris 409460	15	22	21	16
Cenchrus ciliaris 443	12	13	18	12
Means	13	15	21	14

Obtained from: Garcia-Dessommes et al. (2003ab).

The 20% of the 84 new genotypes listed in Table 9.20 had insufficient Zn to meet the needs of adult ruminants. Furthermore, 23 % resulted marginally sufficient. Therefore, the Zn can be limiting for ruminants consuming the new genotypes of C. ciliaris, under rainfed conditions in the Teran county of the state of Nuevo Leon, Mexico.

Zinc deficiency causes a type of leaf discoloration called chlorosis, which causes the tissue between the veins to turn yellow while the veins remain green. Chlorosis in zinc deficiency usually affects the base of the leaf near the stem. Chlorosis appears on the lower leaves first, and then gradually moves up the plant. In severe cases, the upper leaves become chlorotic and the lower leaves turn brown or purple and die.

Table 9.20. Zinc content (mg kg^{-1}; dry matter) in 84 new genotypes of Cenchrus ciliaris collected in Teran county of the state of Nuevo Leon, Mexico in November 2000

Genotypes	Zn, mg kg^{-1}	Genotypes	Zn, mg kg^{-1}	Genotypes	Zn, mg kg^{-1}	Genotypes	Zn, mg kg^{-1}
202513	35	409185	23	409263	34	409459	32
253261	42	409197	35	409264	37	409460	18
307622	30	409200	31	409266	27	409465	35
364428	13	409219	37	409267	38	409466	27
364439	30	409220	37	409269	36	409472	24
364445	17	409222	32	409270	32	409480	11
365654	16	409223	42	409272	34	409529	18
365702	20	409225	33	409273	22	409691	14
365704	23	409227	33	409275	22	409711	18
365713	18	409228	35	409276	19	414447	15
365728	20	409229	33	409278	20	414451	21
365731	28	409230	35	409279	37	414454	20
409142	25	409232	40	409281	38	414460	60
409151	26	409234	40	409282	28	414467	22
409154	14	409235	41	409284	34	414499	21
409155	16	409238	34	409285	34	414511	15
409157	22	409240	37	409287	18	414512	20
409162	18	409242	23	409288	26	414520	22
409164	19	409252	36	409290	28	414532	22

409165	33	409254	34	409291	36	443	36
409168	17	409258	44	409293	45	Nueces	33
Mean of all genotypes							28

Obtained from: Morales-Rodríguez et al., 2005b

It has been reported that the leaves had more Zn in stems than in leaves of the cultivated grasses *C. ciliaris, C. dactylon, D. annulatum* and *P. coloratum* collected in Linares county of the state of Nuevo Leon, Mexico under rainfed conditions (Figure 9.8). Furthermore, there is a clear difference in Zn content between seasons of the year.

Different enzyme systems in the body require mineral zinc as cofactor. These enzyme systems are responsible for every major physiological function that necessitates catalytic activity from enzyme at the molecular level. It has been reported that Zn concentration in the stems was twice higher than of the leaves in most cultivated grasses in US. Therefore, stems in cultivated grasses are an important dietary intake of Zn for grazing ruminants.

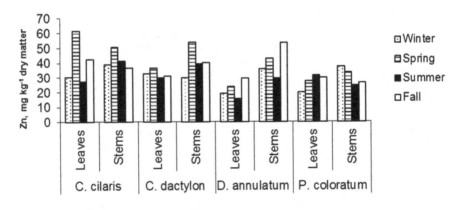

Grasses and parts

Figure 9.8. Seasonal Zn content (mg kg-1 dry matter) in leaves and stems of four cultivated grasses collected in Linares county of the state of Nuevo Leon, Mexico in 1998 and 1999
Obtained from: Ramirez et al. (2003); Ramirez-Lozano (2003); Ramirez et al. (2005a).

Zinc content in native grasses

Native grasses from northeastern Mexico contained more Zn (Tables 9.21 and 9.22) than cultivated grasses growing in the same region (Tables 9.18, 9.19 and 9.20). Furthermore, Zn content in all native grasses followed a clear seasonal tendency. In general, grasses had more Zn in the wet seasons (summer and fall) than in the dry (winter and spring). Moreover, all the grasses and in all seasons contain sufficient Zn to meet the requirements of growing beef cattle (30 mg kg^{-1} dry matter of the diet), with exception of *Bouteloua trifida* and *Panicum hallii* (Table 9.22) that resulted marginal insufficient, except in summer, to meet the needs of Zn for growing ruminants.

Table 9.21. Seasonal Zn content (g kg^{-1} dry matter) in native grasses collected in Marin county of the state of Nuevo Leon, Mexico in 1994

| Native grasses | Seasons | | | | Annual |
	Winter	Spring	Summer	Fall	mean
Aristida longiseta	46	48	52	55	50
Bouteloua gracilis	36	45	59	56	49
Cenchrus incertus	46	45	65	76	58
Hilaria belangeri	39	42	57	61	49
Panicum hallii	59	65	68	68	65
Setaria macrostachya	43	57	66	74	60
Seasonal means	45	50	61	65	55

Obtained from: Ramirez *et al.* (2004)

Table 9.22. Seasonal Zn content (g kg-1 dry matter) in native grasses collected in Teran county of the state of Nuevo Leon, Mexico in 2001 and 2002

| Native grasses | Seasons | | | | Annual |
	Winter	Spring	Summer	Fall	mean
Bouteloua curtipendula	31	45	64	50	48
Bouteloua trifida	15	25	31	27	25
Brachiaria fasciculata	46	45	64	55	52
Digitaria insulares	31	51	46	45	43
Chloris ciliata	61	47	74	60	60
Leptochloa filiformis	49	43	62	61	53
Panicum hallii	29	29	58	29	37
Panicum obtusum	32	44	58	52	46
Panicum unispicatum	66	53	82	76	69
Setaria grisebachii	55	57	69	59	60

Setaria macrostachya	59	59	57	43	55
Tridens eragrostoides	50	64	66	53	58
Tridens muticus	29	38	31	37	34
Seasonal means	43	46	59	50	49

Obtained from: Cobio-Nagao (2004).

Cobalt (Co)

Cobalt is essential for growth of the rhizobium, the specific bacteria involved in legume nodulation and fixation of atmospheric nitrogen into amino acids and proteins in legumes. Vitamin B_{12} that contains Co is synthesized by the rhizobium and circulated in hemoglobin. The hemoglobin content in the nodules is directly related to nitrogen fixation. Thus, a deficiency in Co is shown in reduced Vitamin B_{12} production and lower nitrogen fixation.

Ruminants can use elemental Co; however, the microbes fermenting and digesting plant material in the rumen convert elemental cobalt into vitamin B_{12}, which the animal can use. When ruminants are on a cobalt deficient diet, there is a gradual loss of appetite, weight loss, muscle wasting, depraved appetite, anemia, and eventually death. The animals appear as if they have been starved, except that the visible mucus membranes are blanched and the skin is pale and fragile. Secondary signs of a cobalt deficiency include fatty liver, increased mortality of offspring shortly after birth, increased susceptibility to infectious agents and infertility.

The rapid loss of appetite in cobalt deficient ruminants is not nearly as obvious in vitamin B_{12} deficient monogastric animals. Monogastric energy metabolism is based on glucose absorbed from the small intestine, while ruminants get approximately 70% of their metabolizable energy from volatile fatty acids produced in the rumen. Acetate, propionate and butyrate are the main volatile fatty acids utilized for energy. Normal propionate metabolism requires vitamin B_{12}. Accumulation of propionate in the blood rapidly depresses appetite, and there is an inverse relationship between feed intake and propionate clearance in Co-deficient sheep.

Grasslands containing 0.1 ppm Co will prevent deficiency symptoms, while levels of 0.05 to 0.07 ppm Co are deficient. Further evidence was if 0.1 ppm Co in the diet will meet both the Co and the vitamin B_{12} requirements of sheep; in addition, it was reported that ruminal concentrations of both vitamin B_{12} and propionic acid were maximized at 0.1 ppm dietary Co. However, cobalt concentration of 0.10 to 0.15 mg per kg (0.25 to 0.034 mg per kg) of dietary dry matter resulted in adequate vitamin B_{12} production to meet the requirements of ruminal microorganisms fed a high-concentrate diet in continuous-flow fermenters.

Selenium (Se)

Selenium play a role in antioxidative mechanisms in plants. It act as an antioxidant, inhibiting lipid peroxidation in ryegrass (*Lolium perenne*). In plants, it occurs as a bystander mineral, sometimes in toxic proportions in forage (some plants may accumulate Se as a defense against being eaten by animals, but other plants such as locoweed require Se, and their growth indicates the presence of Se in soil).

Selenium is a necessary mineral for both humans and animals, and naturally occurs in varying concentrations within the soil in all parts of the world. Available selenium in the soil is taken up by many plants, including some cereal grains and forages used for livestock feed. Selenium is an integral component of the glutathione peroxidase enzyme, which works as a cellular antioxidant. The major role of this enzyme is to protect cellular membranes from damage by converting hydrogen peroxide to water. Hydrogen peroxide and other intermediates of cellular reduction pathways can damage cellular membranes, disrupt cellular function and may negatively affect animal health. Tissue concentrations of selenium are associated with glutathione peroxidase activity and are directly related to dietary selenium intake. The liver and glandular tissues have the greatest selenium concentrations of all the bodily tissues. Selenium and vitamin E are often discussed together as they are both involved in protecting cells against free radicals that cause cellular damage. Selenium contributes to the destruction of oxidizing components, while vitamin E prevents their formation and often a deficiency in one leads to an increased requirement for the other. More recently, it has been shown that selenoproteins have roles in immune function and thyroid hormone metabolism.

Selenium has been identified as an essential part of normal reproductive and immune function. Some general indicators of selenium deficiency in ruminants include reproductive failure, retained placenta in dairy cows, increased calf mortality, decreased calf weaning weights, and immune suppression. Selenium deficiency and its relationship to embryonic loss and reduced fertility appear to be more prominent in sheep compared to cattle. Supplementing dairy cows that are deficient in selenium has been shown to reduce the incidence of retained placentas. It is hypothesized that selenium deficiency resulting in reduced glutathione peroxidase enzyme activity combined with a deficiency of vitamin E could allow for the buildup of free radicals, causing damage to cells and tissues and subsequent retention of placenta. Furthermore, selenium readily passes through the placenta therefore the selenium status of the gestating dam will be directly related to the selenium status of the calf at birth. Little data exists to indicate that dietary intake exceeding the 0.30 ppm requirement for dairy cows will result in a reduction of mastitis, however, deficiencies of selenium have been associated with decreased intracellular kill of pathogens by neutrophils, and subsequently decreased immune function. White muscle disease is the most severe clinical sign of selenium deficiency in ruminants and other animals. Some indicators of white muscle disease as most often seen in young calves (or lambs) include:

1. Leg weakness and stiffness
2. Flexion of the hock joints
3. Muscular tremors
4. Chalky striations and necrosis of cardiac and skeletal muscles
5. Death from cardiac failure

Iodine (I)

Iodine is an essential trace element for life, the heaviest element commonly needed by living organisms, and the second-heaviest known to be used by any form of life. The effect of iodine varies depending on plant species. It is not an essential element but added to media to improve growth of roots and callus in tissue culture. The ocean is a rich and natural reservoir of iodine; seafood and seaweed are therefore the richest natural food sources of iodine available.

The main role of iodine in animal biology is as constituents of the thyroid hormones, thyroxine (T4) and triiodothyronine (T3). These are made from addition condensation products of the amino acid tyrosine, and are stored prior to release in an iodine-containing protein called thyroglobulin. The T4 and T3 contain four and three atoms of iodine per molecule, respectively. The thyroid gland actively absorbs iodine from the blood to make and release these hormones into the blood, actions, which are regulated by a second hormone TSH from the pituitary. Thyroid hormones are phylogenetically very old molecules which are synthesized by most multicellular organisms, and which even have some effect on unicellular organisms. Thyroid hormones play a basic role in biology, acting on gene transcription to regulate the basal metabolic rate. The total deficiency of thyroid hormones can reduce basal metabolic rate up to 50%, while in excessive production of thyroid hormones the basal metabolic rate can be increased by 100%. The T4 acts largely as a precursor to T3, which is (with minor exceptions) the biologically active hormone.

Iodine has a nutritional relationship with selenium. A family of selenium-dependent enzymes called deiodinases converts T4 to T3 (the active hormone) by removing an iodine atom from the outer tyrosine ring. These enzymes also convert T4 to reverse T3 (rT3) by removing an inner ring iodine atom; and convert T3 to 3,3'-Diiodothyronine (T2) also by removing an inner ring atom. Both of the latter are inactivated hormones, which are ready for disposal and have essentially no biological effects. A family of non-selenium dependent enzymes then further deiodinates the products of these reactions.

Selenium also plays a very important role in the production of glutathione, a most powerful antioxidant. During the production of thyroid hormones, hydrogen peroxide is produced, high Iodine in the absence of selenium destroys the thyroid gland (often felt as a sore throat feeling), and the peroxides are neutralized through the production of glutathione from selenium. In turn, an excess of selenium increases demand for iodine, and deficiency will result when a diet is high in selenium and low in iodine.

CHAPTER 10

Grazing

Introduction

Grazing is a method of agriculture in which domestic livestock are used to convert grass and other forage into meat, milk and other products. The purpose of a grazing system is to increase forage production and the production of livestock and wildlife, while maintaining or improving the condition of the pasture. Its importance lies in giving a better use of the different forage plants, providing sufficient rest time for that after each grazing period recovered in order to increase the density and forage production in grasses and forbs, mainly in overgrazed areas, in order to allow plants to produce seed and to ensure their propagation. For a grazing system to be effective, it has be based on plant physiology, adapt to the type of plants present, and familiar to the topography of the terrain. It has to improve the production and use of animal feed and increase profit. It must be flexible, economical and of practical application, suitable for the type and class of livestock and has to be in accordance with the objectives of livestock management.

Ecological benefits of grazing

Grazing is favorable to the ecosystem. It is helpful to the soil and grasses, stimulating nutrient dense soil and motivating the growth of plant varieties. During grazing, livestock promotes plant growth, therefore increasing forage production. Moreover, the urine and feces of the animal recycle nitrogen, phosphorus, potassium and other plant nutrients and return them to the soil. It also acts as food for insects and organisms found within the soil. These organisms aid in carbon sequestration and water filtration. Nutrients and organisms, all of which are necessary for soil to be wealthy and capable for production. In addition, grazing helps to promote the growth of native plants and grasses. By livestock grazing, the cultivated grasses are controlled and the native plants can redeveloped. As well as using grazing to increase plant growth, the actual hoof action of the livestock also promotes growth. The trampling helps to insert the seeds into the soil so that the plants and grasses can germinate. On the other hand, grazing can also allow for accumulation of litter (horizontal residue) helping to eliminate soil erosion. Soil erosion is important to minimize because with the soil erosion comes a loss of nutrients and the topsoil. All of which are important in the regrowth of vegetation.

It appears that grazing may stimulate biodiversity. Many species are dependent on ranch grasslands and grazing animals to maintain their habitat. The grasses that are stimulated through grazing provide a habitat for many species. When the land is left unattended or is not grazed, grasses will die with the seasons and accumulate as litter on the ground. For many birds, this is not attractive and they avoid making a nesting area of it. However, when the grass is grazed, the dead litter grass is reduced and allows the birds to use it, while at the same time the livestock benefit. Just as significantly, it increases species richness. When grazing is not used, many of the same grasses grow, therefore making a monoculture.

Grazing systems

It seems that suitable land use and grazing management techniques need to balance maintaining forage and livestock production, while still maintaining biodiversity and ecosystem services. By the use of grazing systems and being sure to allow adequate recovery periods for regrowth, both the livestock and ecosystem may benefit. Beside with recovery periods, producers can keep a low density on a pasture, so as not to overgraze. Controlled burning of the land can be valuable in the regrowth of native plants, and new lush growth. Moreover, producers can increase plant and species richness through grazing, by providing an adequate habitat. Even though, grazing may be difficult for the ecosystem at times, it is clear that well-managed grazing techniques may reverse damage and improve the grassland.

In the 19th century, grazing techniques were virtually nonexistent. Pastures were grazed for long periods, with no rest in between. This led to overgrazing and it was detrimental to the grassland, wildlife, and livestock producers. Today, ranchers have developed grazing systems to help improve the forage production for livestock, while still being beneficial to the land.

Continuous and season-long grazing

These systems are technically not grazing systems per se because there is no try to leave a portion of the range ungrazed by livestock for at least part of the growing season. These grazing systems work best on flat, well-watered areas (e.g., watering points no more than 4 km apart) where precipitation occurs as some light rains throughout the summer, and where most plants have some grazing value.

Seasonal

Seasonal grazing incorporates grazing animals on a particular area for only part of the year. This allows the land that is not being grazed to rest and allow new forage to grow. Dissimilar vegetation types are typically fenced; however, livestock movements can also be controlled by turning on (or off) watering points, a technique most commonly employed in the Southwestern U.S. and Northern Mexico.

Best Pasture

This system tries to tie cattle movements with irregular rainfall trends and associated forage production without respect to a rigid rotation schedule. For example, when a local rain event causes a flush of annual forbs in a particular pasture, cattle are moved to that pasture, and then moved back to the previous pasture once acceptable utilization levels of the temporary forb resource have been reached. Instead, if a pasture that is tentatively scheduled for grazing continues to miss localized rainstorms while another pasture continues to receive moisture, the rotation schedule for the two pastures could be drooped. Because livestock movements are not strictly timed to a particular schedule, the best pasture system requires that land managers command an approach of great plasticity. As with the seasonal grazing system, the best pasture system may involve turning on (or shutting off) watering points in grazed (deferred or rested) pastures.

Rotational

Rotational grazing involves dividing the range into several pastures and then grazing each in sequence throughout the grazing period (Figure 10.1). Utilizing rotational grazing can improve livestock distribution while incorporating rest period for new forage. However, experimental evidence indicates that rotational grazing is a viable grazing strategy on rangelands, but the perception that it is superior to continuous grazing is not supported by the vast majority of experimental investigations. There is no consistent or overwhelming evidence demonstrating that rotational grazing simulates ecological processes to enhance plant and animal production compared to that of continuous grazing on rangelands.

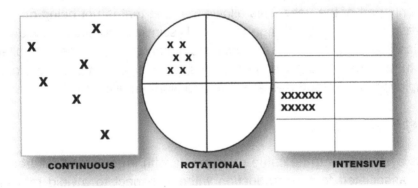

Figure 10.1 Continuous, rotational and intensive grazing systems

Rest rotation

Rest rotation grazing divides the range into at least four pastures. One pasture remains rested throughout the year and grazing is rotated amongst the residual pastures. This grazing system can be especially beneficial when using sensitive grass that requires time for rest and regrowth. For example, one pasture in a 3-year, 3-pasture rest-rotation might be managed as follows during a 3-year cycle: 1) Graze the entire year or growing season, 2) Defer, then graze, and 3) Rest. This schedule rests about 1/3 of the range annually. Rest provides an opportunity for the vegetation around natural or developed water to recover and helps meet multiple use objectives (e.g., providing hiding cover for birds and mammals, leaving ungrazed areas for public viewing and enjoyment).

Deferred rotation

Deferred rotation involves at least two pastures with one not grazed until after seed-set. By using deferred rotation, grasses can achieve maximum growth during the period when no grazing occurs. Several modifications of this system have been used including more than two pastures; however, its important feature is that each pasture periodically receives postponement (usually every 2 to 4 years, depending on the number of pastures).

Santa Rita

This system is a one-herd, three-pasture, three-year, rest-rotation system that was modified for midsummer rainfall and concomitant forage production patterns that typically occur in the hot semi desert grasslands. A three-year rotational schedule for one pasture is as follows: 1) Rest 12 months (November to October), 2) Graze 4 months (November to February), 3) Rest 12 months (March to February), and 4) Graze 8 months (March to October). Each pasture receives rest during both early spring and "summer-monsoon" growing periods for 2 out of every 3 years, but each yearly forage production is also grazed (first growth of the years is grazed in winter). A full year of rest before spring grazing allows residual vegetation to accumulate which helps protect new spring forage from heavy grazing. Target utilization levels in grazed pastures are 30-40%. It was argued that the Santa Rita system promoted recovery of ranges in poor condition, but had little advantage over moderate continuous grazing on ranges in good condition.

Patch-burn grazing

This system burns a third of a pasture each year, no matter the size of the pasture. This burned patch attracts the grazers (cattle or wildlife) which graze the area heavily because of the fresh grasses that grow in. The other patches receive little to no grazing. During the next two years, the next two patches burn consecutively and then the cycle begins again. In this way, patches receive two years of rest and recovery from the heavy grazing. All this results in a diversity of habitats that different prairie plants and birds can utilize, mimicking the effects of the pre-historical bison/fire relationship where bison heavily graze one area and other areas have opportunity to rest.

Riparian area grazing management

Riparian area grazing is used more towards improving wildlife and their habitats. It uses fencing to keep livestock off ranges near streams or water areas until after wildlife or waterfowl periods, or limiting the amount of grazing to a short period. A riparian area is defined as the strip of moisture-loving vegetation growing along the edge of a natural water body. The exact boundary of the riparian area is often difficult to determine because it is a zone of transition between the water body and the upland vegetation. A riparian management zone usually extends from the edge of water to the upland area.

Natural riparian vegetation usually has deep roots. The deep root mass helps maintain the bank or shoreline structure by holding the soil together. In addition, riparian vegetation can also help reduce the amount of sediment and nutrients that are transported in runoff. The vegetation physically traps sediment in surface flow, and uses the nutrients in the shallow sub-surface flow. Moreover, riparian vegetation provides shade. Shade helps regulate stream temperatures by controlling the amount of sunlight that reaches the stream. Furthermore, riparian vegetation is a source of small organic debris, which may include leaves, twigs and terrestrial insects. Riparian vegetation helps reduce stream velocity during high flow events.

Implications of grazing systems

Applying any grazing system does not exclude the need to note basic principles of grazing management (e.g. stocking rates, season of use, kind or mix of animals, animal selectivity and others). Grazing systems require more, rather than less management effort than to continuous or season-long grazing. Augmented attention to range and livestock management may frequently be a first reason for the triumph of a specific grazing system. Animal distribution tools such as riding, suitable location of nutrient blocks, discriminatory removing based on animal behavior characteristics, range enhancements (burning, reseeding, water improvements), and control of access to watering locations should be applied in a manner that tie the planned management commitments of grazing systems. Flexibility is the guarantee of successful range management in arid regions. Exacting observance to animal numbers and livestock movement dates without regard to vagaries in precipitation and forage production may be counterproductive to both rangeland and livestock production. Adjust stocking rates and rotation dates in a way that livestock numbers are in balance with forage supply. Monitoring the rangeland is very important to certificate both successes and failures of grazing systems and other management actions. Rangelands are exceptionally variable in the kind and amount of vegetation that they are able of generating. This variability is apparent across the land (space) and across the years (time). Monitoring techniques are available to help you determine how much variability you can expect on your ranch across both space and time.

Mixed species grazing system

Mixed species grazing is when two or more species of domestic animals are grazed together or separately on the same grazing area in the same grazing season (Figure 10.2). The rationale for mixed species grazing is based on the principle that animals have different grazing preferences and dietary overlap is minimal in a diverse grassland. The productivity of the mixed species system is usually better than the system of that uses one species. However, mixed economic performance of the system will depend on productivity in terms of milk and meat production and the price of the products. Intake and diet composition are the most important factors that define the productivity of grazing animals. Intake and diet composition of animals depend on the characteristics of the pasture, the type of animals and management factors, including the stocking rate, so the above factors must be taken into account for a system of mixed species grazing.

Figure 10.2. Sheep and goats grazing in the same pasture

Fundamentals of mixed species grazing system

Using mixed species grazing system is about ensuring a more efficient use of feed resources and thus, increase the productivity of rangeland (Figure 10.3). An additional advantage is that the system increases the diversity of animal species; the risk of loss is reduced. In most experiments on mixed species grazing, the total animal production per unit area has been increased with mixed species grazing than grazing using one animal species. The augmented production of mixed species system is based through analysis and selective grazing habits of diet, stocking rate, relationship between species and animal equivalent characteristic of the pasture, parasites and social factors.

Several factors may influence the application of the mixed species grazing system:

1. Small herbivores generally require more energy relative to their gut capacity than large ones and thus have to select higher quality foods. Conversely, larger animals with comparatively large gut capacity in relation to their metabolic needs may retain digesta in the gastrointestinal tract for extensive and thus digest it more exhaustively. The physiological state of ruminants might also affect its dietary choice.

2. Species effects on selectivity and thus on biodiversity are of great importance and are among the better understood of the effects of grazing animal type. Ruminant species have been classified into three main feeding categories, grazers, intermediate feeders and browsers and much literature has accumulated on the morphological and physiological adaptations, which allow animals in the different categories to efficiently extract the nutrients from the diet consumed.

Figure 10.3. Goats and sheep consuming forage in semiarid land

3. The use of traditional or rustic livestock breeds is often recommended for nature conservation management. While such recommendations are partly based on the perceived

hardiness of these animals it is often hypothesized that the use of commercial breed types may pose a risk to the efficient stability of biodiverse grassland communities through hard-adapted animal behavioral responses.

4. Reports of breed differences in selectivity and impact on grass structure and biodiversity, suffer from a failure to separate true genetic differences between breeds from environmental effects, particularly previous experience of biodiverse pastures during early life that may affect subsequent selection. It has been evidenced than shortly exposing animals to new plant species at a young age disturbs their successive grazing selections.

5. Dependable with the important effect of body size on dietary choice, young animals and females usually show greater selectivity than older animals or males. It was determined that the selectivity of calves was greater than that of older cattle, based on that the nitrogen concentration in their feces was higher and the fiber concentration was lower; in addition, the C33 alkane concentration in the faces was higher demonstrating that more leaf material was consumed.

Differences in grazing habits

Competition for forage. There is considerable information indicating that grazing sheep, goats and cattle differ in their preferences for consuming certain species or plant parts (Table 10.1). In mixed grazing of sheep, goats and cattle, goats selected a diet higher in native shrubs, sheep also selected shrubs, but they have preferred grasses and cattle eat mainly grasses. Usually, the diet of goats is higher in crude protein and organic matter digestible and has less fiber than the diet selected by sheep or cattle.

Table 10.1. Differences in diet preferences and strategies of grazing by ruminants

Type of plants	Cattle	Sheep	Goats
	Percentage in diet		
Grasses	70	60	20
Weds	20	30	20
Shrubs/forbs	10	10	60

Strategies of grazing	Strategies of grazing		
	Mixed (grasses)	Mixed (shrubs)	Mixed
Selectivity index	High	High	Intermedium
Digestion of fiber	High	Low	Intermedium
Grasses/shrubs	Intermedium	High	High

In rangeland, with high plant diversity and stocking rate is too low, there is no competition for the plant species between goats and sheep and between goats and cattle, but there is between sheep and cattle. However, when sheep and cattle graze on cultivated grasses, where the diversity of plant species is low, differences between diets may not exist. Sheep can graze forage close to the ground when the forage available tends to be limited, so are more competitive than cattle when stocking rate is high.

Advantages and disadvantages of mixed species grazing system

The mixed grazing systems using sheep, goats and cattle has advantages and disadvantages, depending on the particular conditions of the farms where the system is practiced. The advantages may be: 1) guarding sheep from predators, 2) eating plants by sheep and goats that cattle do not consume, 3) increasing the stocking rate, 4) diversifying income, 5) may be consuming meat in the ranch and 6) controlling of parasites. However, there may be disadvantages such as: 1) it is necessary to exercise extreme health measures, 2) is required to supplement nutritional deficiencies for each animal species, 3) is need to build economic fenced, 4) general management problems are presented and 5) there is a lack of tradition for the implementation of mixed grazing. Therefore, by controlling the distribution of animals in mixed grazing in pastures, it is possible to increase the quality of forage resources and a better comprehensive productivity of animals with a consequent increase in meat or milk production and economic benefits.

Overgrazing

Overgrazing (Figure 10.4) may happen under continuous or rotational grazing systems. It may be produced by having too many animals on the pasture or by not properly supervisory their grazing activities. Overgrazing reduces plant leaf areas, which diminishes capture of sunlight and plant growth. Plants become debilitated and reduce root length, and the pasture turf declines. The diminished root length creates the plants more vulnerable to death during droughts. The debilitated grass permits weed seeds to germinate and grow. The weeds may be unpleasant or toxic.

One indicator of overgrazing is that the animals track little of pasture. Under continuous grazing, overgrazed pastures are predominated by short-grass species. Soil may be noticeable between plants in the stand, allowing erosion to occur. Under rotational grazing, overgrazed plants do not have enough time to grow to the proper height between grazing events. Another indicator is that the livestock run out of pasture, and hay needs to be fed early in the fall. The potential grazing season continues into November or longer when winter grazing management is implemented. If hay feeding is needed in October under normal weather conditions, the pasture probably is being overgrazed. Overgrazing is also indicated in livestock performance and condition.

Figure 10.4 Photo that show overgrazed pasture

Cows having inadequate pasture in the early fall do not have a chance to gain weight after the calves are weaned and may have poor body condition going into the winter. This makes them hard to winter and may reduce the health and vigor of cows and calves at calving. Moreover, cows in poor body condition do not cycle as soon after calving, which can result in delayed breeding. Overgrazing may increase soil erosion. Reduced soil depth, soil organic matter, and soil fertility. Soil fertility can be corrected by applying the appropriate lime and fertilizers. However, the loss of soil depth and organic matter requires years to correct. To prevent overgrazing, match the forage supplement to the requirement of the herd. This means that a buffer needs to be in the system to adjust for the fast spring growth of cool-season forages.

CHAPTER 11

Edible trees and shrubs in pastures

Introduction

It has been documented that the future of livestock should include forest-pastoral (silvopastoral) systems, which include shrubs and trees with edible leaves or fruits as well as herbage. Consumers are now demanding more sustainable and ethically sourced food, including production without negative impacts on animal welfare, the environment and the livelihood of poor producers. Silvopastoral systems discourse all of these aspects with the added benefit of increased production in the long-term basis. As a sustainable alternative, investigators promoter the use of a diverse group of edible plants such as that in a silvopastral landscape, which promotes: healthy soil with better water retention, helps fight soil erosion, encourages predation of harmful animals, minimizes greenhouse emissions, reduces the carbon footprint of the activity and minimizes greenhouse emissions, reduces injury risk and stress for animals, increases biodiversity and improves job satisfaction for farms. The planting as forage plants of both shrubs and trees whose leaves and small branches can be consumed by ruminant animals can transform the prospects of obtaining sustainable animal production. Such planting of fodder trees has already been successful in several countries. It is clear that silvopastoral systems increase biodiversity, improve animal welfare and provide good working conditions while enabling a profitable farming business. The next step is to get farmers to adopt this proven, sustainable model.

Trees and shrubs

When ruminants, such as cows, goats and sheep, are consuming the plants from a silvopastoral system, researchers have seen an increase in growth and milk production. Milk production in the tropical silvopastoral system mentioned above was 4.13 kg per cow when compared with 3.5 kg per day on pasture-only systems. As the numbers of animals per hectare was much greater, production of good quality, milk per hectare was four to five times greater on the silvopastoral system. One of the additional benefits of using the silvopastoral system is that it increases biodiversity. Biodiversity is declining across the globe, and the main culprit is farming 33% of the total land surface of the world is used for livestock production. If farmers were to switch to sustainable livestock production methods, such as the silvopastoral system, the result would be much greater biodiversity with no increase in land use.

Trees and shrubs, often called browse or topfeed, have long been considered important for the nutrition of grazing animals in some countries, particularly in those areas with a marked dry season (Figure 11.1). They provide a supplement of green fodder when grasses and other herbaceous material is dry and they provide the only source of protein and energy during drought when all other feed is absent. However, trees and shrubs have several disadvantages as sources of fodder. They are often inaccessible to grazing animals. They are slow to establish requiring isolation from stock. Their foliage generally has higher fiber and lignin content than grasses, and often has higher levels of tannins and other secondary compounds than shorter-lived herbaceous plants. Although sometimes higher in protein, they often have lower energy value than herbaceous plants due to their lower digestibility.

An accurate assessment of the forage value of browse species can only be made from the response of grazing animals. A summary of the factors important in assessing forage value.

1. Economic value. Assuming that a browse plant can meet the required forage value. Its economic value to the

property-owner will depend on some additional characteristics of the plant, the cost of establishing and managing the trees and shrubs, the time taken for them to reach a productive age, the persistence of the plant under regular use, and the ability of the plant to produce fodder when it is most needed.

2. Landscape values. Like all trees and shrubs, browse species can make a valuable contribution to landscape stability by decreasing the risk of wind and water erosion. They also contribute to the cycling of nutrients, especially nitrogen in the case of legumes, and the cycling of water. Their deeper and more permanent root systems also help to overcome problems of declining surface structure and compaction commonly associated with crop and pasture land. The introduction of browse plants can also increase the genetic diversity in the landscape and therefore help to increase the resilience of the vegetative cover in general to pests, diseases, fire and climatic extremes. The use of perennial plants can also increase the energy efficiency of agriculture by reducing the requirement for cultivation; in addition, in the case of legumes it may be reduced the use of fertilizer.

3. Distribution of species by climatic zones. Browse plants have some role in grazing productions in all of these zones. Native species are most important in the semi-arid and arid zones; while, cultivated species are being used in the subtropical and temperate zones.

Figure 11.1 Photo of a cow on a *Cenchrus ciliaris* pasture consuming browse fodder

The ideal browse plants

Trees and shrubs with potential have to demonstrate good growth characteristics: many browse plants fit well within these standards; cause minimal soil disturbance because of edible trees need no soil disturbance after establishment; avoid nutrient loss from the soils due to their durability, browse plants help to prevent nutrient losses; have little or no need for irrigation because of browse plants can tap deep water tables out of the reach of short-rooted pastures; some species in dry or marginal areas may need a little irrigation only until they are established; provide yields of edible components comparable to pastures because of most trees and shrubs suitable for fodder have yields reported to be several times higher than pastures grown in comparable areas; have the ability to fix nitrogen to reduce the dependence on artificial fertilizers: most legumes, such as tagasaste, and some other species such as casuarinas, fix atmospheric nitrogen by use of bacteria and convert it into nitrates that the plants can use, removing all dependence on nitrogenous fertilizers; have a high protein content in their edible portions because as many of the trees and shrubs suitable for fodder are legumes and the protein content is high (20 to 50%).

Some of the species that fulfill these minimum requirements include Tagasaste: *Chamaecytisus proliferus,* Carob: *Ceratonia silique,* Honey Locust: *Gleditsia triacanthos,* Willows: *Salix* spp, especially the weeping willow, *Salix babylonica,* and the hybrid *Salix matsudana* x *alba,* Poplars: *Populus* spp, Leucaena: *Leucaena leucocephala* and Chenopods: particularly *Atriplex nummularia.*

Role of browse plants in the nutrition of ruminants

With browse plants in ruminant diets can be seen as three ways:

1. As a N and mineral supplement to enhance fermentative digestion and microbial growth efficiency in the rumen of cattle on poor quality forage.
2. As a source of postruminal protein for digestion. In this role, the influence of secondary plant compounds in binding protein and making it insoluble is of particular importance.

3. As a total feed, supplying almost all the biomass and other nutrients needed to support high levels of animal production.

A further role is presently emerging, that is the potential of constituent secondary plant compounds to manipulate the balance of microorganisms in the rumen. The ability of secondary plant compounds to reduce protozoal populations has important implications for protein availability to ruminants.

Browse plants constitute a major source of food for goats, white-tailed deer and sheep in northeastern Mexico. It has been found that for grazing ruminants, browse is complementary to grasses, especially during the non-growing season, when it provides essential protein when grasses are dormant. Quality and quantity of forage intake by grazing animals is important for the identification of the constraints to efficient production of livestock on rangelands where low intake is the most common factor.

Restriction in nutrient intake is probably the main factor that limits performance by grazing animals. In ruminants however, intake is closely related to both the nutritive value and dry matter availability of forage resources. On open rangelands, the quality and quantity of forage varies appreciably with climate and often leads to nutritional inadequacy. The use of forages from shrubs and trees as supplements can be an alternative strategy that perhaps has not been given adequate research and development attention. For the high nutritive value (protein, minerals and vitamins) and low cost this approach, more than any other, has tremendous potential for application with ruminants, especially where animals are abundant and are managed under extensive conditions.

Cenchrus ciliaris pastures mixed with small trees and shrubs

The nutritive value and dry matter availability of forage are closely related to efficiency of animal production. However, due to the dynamic nature of vegetation, the provision of adequate forage for grazing sheep poses a challenge for range nutritionists. One cannot master this challenge without a detailed knowledge of the physical characteristics of the pasture and the nutritional requirements of sheep for growth and production. Common buffelgrass (*Cenchrus ciliaris*) is one of the most valuable seeded grasses on rangelands of southern Texas, and northern Mexico (Figure 11.2). It has been found that buffelgrass pastures increased carrying capacity, but not necessarily better nutrition, than a mixed-brush range. Native browse plants can be an important nutrient source on buffelgrass pastures as well as forbs, because buffelgrass foliage decreases in quality with age (maturity) and season.

Figure 11.2. Photo of a *Cenchrus ciliaris* pasture mixed
with edible shrubs at northeastern Mexico

Forage selection by sheep

It has been reported that monthly preference indices, during one year, for forage species selected by range sheep grazing on a buffelgrass pasture mixed with native shrubs in northeastern Mexico (Table 11.1). The annual diets of lambs comprised 85.2% grasses, 14.0% shrubs, and 0.8% forbs (Table 11.2). Lambs had heavy consumption of shrubs during spring and end of autumn seasons. *Acacia amentacea* was the shrub species most selected by lambs during the year with 6.0% (annual mean). *Cercidium macrum* was the second most important shrub in lamb diets. However, lambs consumed not only leaves (2.0% annual mean), but also fruits and flowers (1.0%). In these regions, shrubs such as *A. amentacea, C. macrum* and *Porlieria angustifolia* are heavily consumed by goats and white-tailed deer. Other shrubs were observed in lamb diets; however, their contribution was low. Most shrubs that grow in northeastern Mexico are perennial legumes with high crude protein (CP) content. Thus, lambs might have selected browse species to fulfill their protein needs when grass was low in CP during the year.

Grass was the main class of forage eaten by lambs during the year (Table 3). During winter and summer seasons, lambs consumed greater amounts of grasses, but during spring, lambs consumed the lowest amount of grasses, when shifted to shrubs.

Table 11.1. Botanical composition of a buffelgrass (*Cenchrus ciliaris*) mixed with other group of plants

Shrubs	%	Forbs	%	Grasses	%
Acacia amentacea	11.3	Dyssodia acerosa	0.1	Cenchrus ciliaris	70.3
Cercidium macrum	8.3	Coldenia greggii	0.1	Aristida spp.	1.0
Leucophyllum frutescences	3.3	Ruellia corzoi	0.1	Pappophorum bicolor	0.3
Prosopis glandulosa	2.0	Haplopappus spinolosus	0.1	Total grasses	71.7
Acacia farnesiana	1.8	Total forbs	0.6		
Opuntia leptoculis	0.2				
Castela texana	0.3				
Porlieria angustifolia	0.3				
Alicia gratísima	0.4				
Ziziphus obtusifolia	0.1				
Total shrubs	27.7				

The following shrubs: (*Eysenhartia texana, Xanterosilum brachiaria*), forbs (*Solanum eleagniflorum, Sida filicaulis, Palafoxia texana, Oxalis dichandeofolia, Wedelia* sp., *Chamaesaracha sordia, Pathenium* sp., *Hibiscus cardiophylus, Verbena spinoscens, Amaranthus retuflexus*) and grasses (*Panicum hallii*) appeared < 0.1%.

Buffelgrass contributed 85% (annual mean) to the total diet of lambs (Table 11.2). Panicum hallii and Setaria macrostachya were selected by lambs, but in very low and variable amounts during the year. Shrubs such as A. amentacea, C. macrum, L. frutenscens, Prosopis glandulosa, P. angustifolia and Aloysia gratisissima were selected by lambs, but P. angustifolia appeared to be the shrub species more preferred by lambs during all year (Table 11.3). Forb preference indices were inconsistent. Moreover, forbs were consumed by lambs in proportion to their occurrence in pasture at lower ratios. These patterns agree with other research reports. Buffelgrass was the only plant species preferred by lambs during all months.

Table 11.2. Botanical composition of the selected monthly diet by sheep on a *Cenchrus ciliaris* pasture mixed with shrubs and forbs plants

Plant species	Jan.	Feb.	Mar.	Apr.	May.	Jun.	Jul.	Aug.	Sep.	Oct.	Nov.	Dec.
Shrubs												
Acacia amentacea	6.6	1.5	0.5	5.0	17.9	4.5	12.4	1.1	1.7	3.8	5.2	7.5
Cercidium macrum												
Leaves	0.5	0.3	0.0	8.6	5.2	2.5	3.7	3.0	2.1	0.8	2.6	1.4
Fruits and flowers	0.0	0.0	0.0	8.2	0.3	2.0	0.0	0.0	0.0	0.0	0.0	0.0
Porlieria angustifolia	1.3	1.4	0.0	4.5	4.5	1.7	0.9	0.3	0.3	1.5	3.2	1.1
Prosopis glandulosa												
Fruits	0.0	0.0	0.0	4.0	1.4	4.6	0.0	0.0	0.0	0.0	0.0	0.0
Leaves	1.1	0.0	0.2	0.1	0.9	0.6	0.2	0.1	0.3	1.0	0.9	0.9
Leucophyllum frutescences	1.3	0.8	1.1	0.0	0.1	0.5	0.2	0.1	0.0	1.6	2.4	0.8
Cordia boissieri	0.0	0.0	0.0	4.6	0.0	0.0	0.0	0.0	0.0	0.0	0.0	0.0
Total shrubs[a]	11.1	4.4	1.8	36.5	30.5	17.8	18.8	4.7	5.3	9.3	17.1	11.8
Forbs												
Palafoxia texana	0.02	0.0	0.1	0.0	0.2	0.1	0.0	0.4	0.8	0.2	0.1	0.0
Total forbs[a]	0.4	0.1	0.8	0.2	1.3	0.1	0.6	2.2	1.3	1.4	0.8	0.1
Grasses												
Cenchrus ciliaris	88.5	95.5	97.3	63.3	68.3	82.2	80.6	93.1	93.5	89.3	81.8	87.9
Total grasses[a]	88.5	95.5	97.3	63.3	68.3	82.2	80.6	93.1	93.5	89.3	81.8	87.9

[a]Totals include shrubs with less than 0.4% in the annual diet and forbs with less than 0.2%.

The Crude protein content was content of selected diets by range sheep was 13% (annual mean) (Table 11.4). Lambs selected diets with high CP during April, May, June, July and August, compared with other months. Shrubs and forbs seemed to make a significant contribution to CP intake, as the highest proportions of shrubs and forbs in diets (Table 11.2) coincided with the highest levels of dietary CP. It has been found that forbs and shrubs contain higher CP than grasses. Since buffelgrass was the main diet component of lamb diets during the year, and this grass is a warm-season species, temperature and rainfall factors influenced its CP content. It has been reported that CP content of hand harvested common buffelgrass foliage in Texas varied from 3.6% in January to 14.2% in October. Beacuse diets of lambs were never lower than 10% CP, browse species in diets may have maintained adequate levels of CP for lamb production.

Table 11.3. Preference indices by sheep on a *Cenchrus ciliaris* pasture mixed with shrubs and forbs

Species	Jan.	Feb.	Mar.	Apr.	May.	Jun.	Jul.	Aug.	Sep.	Oct.	Nov.	Dec.
Shrubs												
Acacia amentacea	1.1	0.1	0.01	1.9	3.9	0.4	1.2	0.5	0.1	0.4	0.5	1.1
Cercidium macrum	0.1	0.0	0.0	2.0	1.5	0.1	0.4	1.5	0.3	0.0	0.4	0.0
Leucophyllum frutescences	0.2	0.1	0.3	0.0	3.9	0.0	0.0	4.7	0.0	0.9	0.4	1.1
Prosopis glandulosa	7.8	0.0	0.03	0.0	1.6	0.1	4.5	0.0	0.3	0.4	0.2	0.1
Porlieria angustifolia	5.7	5.5	0.1	42.1	31.5	8.0	1.5	0.6	0.3	8.2	33.3	2.9
Aloysia gratisissima	2.5	-	0.7	+	+	+	-	+	0.0	1.2	+	-
Forbs												
Coldenia greggii	0.0	0.5	-	0.0	+	+	+	-	0.0	2.1	1.6	0.0
Palafoxia texana	-	+	-	+	-	-	+	7.2	7.0	-	-	+
Sida filicaulis	3.0	+	+	+	+	+	+	0.0	0.0	+	+	+
Verbena canescens	+	+	+	+	+	+	+	0.0	0.0	+	1.0	0.0
Grasses												
Cenchrus ciliaris	1.2	1.3	1.3	1.0	1.0	1.3	1.3	1.2	1.3	1.3	1.3	1.4

0, means that the species was sampled in the transects, but was not in the diets; -, means that the species was present only in the diets; +, means that the species was not present in the diets neither in the transects.

Annual mean of NDF content was 76.0%. The highest NDF in forage selected by lambs was during May, November and December, and the lowest during January and March (Table 11.4). High levels of grasses in diets may be associated with high levels of NDF. It has been reported that grasses contain much higher levels of NDF than do browse leaves or forbs. Dietary ADF was different among months (46.6% annual mean). The highest level occurred during May (Table 11.4) and the lowest during June. High cell wall levels in forage consumed by lambs may be limiting factors reducing animal intake and digestibility by microbes in the rumen. However, studies in Texas have reported that NDF content of buffelgrass is high and variable during the year, varying from 65% to 78% and ADF from 31.7% to 49.9%.

Table 11.4. Nutrient content and digestibility *in vitro* of extrusas by sheep grazing on *Cenchrus ciliaris* pasture

Items[a]	Jan.	Feb.	Mar.	Apr.	May.	Jun.	Jul.	Aug.	Sep.	Oct.	Nov.	Dec.
Crude protein, %	10.5	12.2	11.9	15.3	14.2	14.8	17.1	16.4	12.8	10.7	10.8	10.6
Neutral detergent fiber, %	68.3	72.3	68.7	77.3	80.3	77.0	77.0	77.7	74.0	77.0	82.1	80.0
Acid detergent fiber, %	48.3	45.1	46.1	47.7	56.4	37.0	40.1	50.0	49.4	46.7	50.6	42.0
IVOMD, %	53.5	61.8	55.9	62.2	64.4	65.1	65.1	64.0	61.4	46.3	45.1	56.5

[a]Dry matter.
IVOMD = in vitro organic matter digestibility.

The *in vitro* organic matter digestibility (IVOMD) of extrusa samples from esophageal fistulated range sheep (Table 11.4). Digestibility data of this study are in agreement with other early studies. It has been reported that IVOMD of buffelgrass plants clipped to a 2-3 cm stubble height, grown at Beeville, Texas, averaged 61.1% during the growing season and 56.7% during the winter dormant season. Moreover, other study reported that buffelgrass plants in Texas had the highest IVOMD values in May and August-September, which coincided with peak rainfall, while October-December was low. These findings agree with higher IVOMD occurred during spring and summer seasons. It has been reported that CP content in buffelgrass influenced the energy value of the digestible OM. Higher IVOMD values coincided with higher CP values of extrusa samples.

Performance of lambs grazing buffelgrass pasture

Successful sheep production on pasture depends on sufficient daily forage intake to meet sheep nutritional requirements; otherwise, deficiencies must be corrected through supplementation if animal performance is to be maximized. It was investigated the influence of the level of energy supplementation on the energy requirements for growth of lambs. Forty recently weaned Pelibuey x Rambouillet lambs (25 castrated males and 15 females), weighing 13.50.8 kg were assigned to each of the five treatment levels of an energy supplement (8/treatment; 5 males and 3 females) at 0.8%, 1.1%, 1.4%, 1.7% and

2.0% of body weight (BW). The supplement offered to lambs provided amounts of energy and protein for a maximum growth of about 147 g per day with a level of energy supplementation of 2.0% BW (Table 11.5). This average daily gain (ADG) may be acceptable for grazing lambs on a buffelgrass pasture. Moreover, the daily metabolizable energy (ME) required (1.50 Mcal) for maintenance and growth for a lamb weighing 25 kg and gaining 100 g day^{-1} was comparable to values reported in the literature. Using sorghum grain as a supplement, only the level of 0.8% BW gave an economical response. The relationship between ADG (g day^{-1}) and digestible energy intake (Mcal kg$^{0.75}$ day^{-1}) was Y = 105:0 + 0:65X (r = 0.67; P<0.05). Daily digestible energy for maintenance was 105.0 kcal DE kg$^{0.75}$, and for gain was 0.65 kcal DE g gain kg$^{0.75}$, and the relationship between ADG (g day^{-1} and ME intake (kcal ME kg$^{0.75}$) was Y = 85:9 + 0:54X (r = 0.67; P<0.05).

Table 11.5. Means of the initial weight (IW), final weight (FW), average daily gain and feed efficiency of lambs fed during the performance test

Concept	Energy supplementation level (% of live weight)					
	0.8	1.1	1.4	1.7	2.0	SE
Initial weigh, kg	14.0	1.3	15.1	15.8	15.8	0.7
Final weight, kg	22.3	22.7	26.6	27.0	28.9	1.1
ADG, g day^{-1}	90.0	110.5	124.6	119.0	147.4	6.5
Feed/gain[a]	7.7	7.7	7.6	7.6	7.6	0.3

ADG = average daily gain.
[a]Intake of organic matter g day^{-1}/daily weigh gain in g.
SE = standard error.

Buffelgrass pastures in northeastern Mexico that grow mixed with native range shrubs may provide a better nutrition for grazing ruminants than pastures with only buffelgrass, especially during spring and summer seasons, and maintain body condition during fall and winter seasons.

Living fence as a source of forage

A living fence (Figure 11.4), often traditionally called a hedgerow, is a permanent hedge tight enough and tough enough to serve any of the functions of a manufactured fence privacy and security, livestock control, protection of crops, but which offers many biological and agricultural services the manufactured fence cannot. A living fence is an excellent example of edge habitat so supportive of ecological diversity on the homestead. As more species find food and habitat there are insects, spiders, toads, snakes, birds and mammals. Depending on species used, living fences can in addition provide food and medicines for people; feedstock for production of biofuels; fodder for livestock. Certainly, appropriate forage trees can be even more productive of livestock feed value than pasture on a per unit basis; and the foliage of some of them contains more protein than that typical protein forage crop and alfalfa hay. Livestock also benefit from the shade of a dense living fence.

If it is included legumes such as Leucaena (Figure 11.3) and other nitrogen-fixing species in the fence, nitrogen will be added to the soil throughout the root zone, and some of that nitrogen can be harvested as well in the form of leafy cuttings for mulches and composts. It has even been argued that a field sheltered by a living fence will retain more CO_2 at ground level, to the benefit of pasture plants or growing crops, which respond with increased yields. In addition, they contribute to buildup of humus in the soil, not only through breakdown of leaf litter, but through root hairs shed to balance loss of top growth each time they are clipped or grazed. If sited along contours, hedges can eliminate rainfall erosion on slopes.

Figure 11.3. Photo of a living fence browsed by cattle

Forage banks

The forage bank are areas dedicated to the production of high quality, nutrient-dense animal fodder. They vary from single strata monocultures to multistrata diverse systems. The reasons for cultivating them vary to some extent from farm to farm and climate to climate. In tropical regions where there is a long and distinct dry season, these forage banks can produce feed in times when there otherwise is not any. In some cases, farmers are using them simply to reduce the amount of feed concentrate they have to purchase. Other farmers have completely eliminated the use of feed concentrate by substituting it entirely with a diverse mix of fodder crops. Often fodder crops are trees and shrubs which are cut back after short durations (one to three months) to ground level, knee height, or sometimes even higher. They can be nitrogen fixers such as many of the trees in the legume family. They can also be other fast growing, easily regenerating, and often highly nutritive species.

Leguminous fodder trees or shrubs (e.g., *Gliricidia sepium, Desmodium rensonii,* etc.) are established as small stands on certain portions of the farm or pasture area serving as a supplementary source of protein-rich fodder for livestock. They also serve as fence. They are regularly cut and the top and branch clipping are then fed to animals.

Cacti as a forage bank

Opuntias used for animal feeding are abundant, easy and cheap to grow, palatable and drought tolerant. Such characteristics make them a potentially important feed supplement for livestock, particularly during periods of drought and low feed availability. A large portion of Opuntia plant biomass is vegetative material rather than fruits, and it can be fed to livestock as fresh forage or stored as silage for later feeding. The idea of using cactus to feed livestock is not recent. It was certain that feeding cactus to livestock started in the USA before the Civil War, and before and after the war; there was extensive freight transportation of cactus pads between Brownsville, Indianola, San Antonio and Eagle Pass in Texas. The plant has become important fodder in many parts of the world, based on natural and cultivated populations. It is cultivated in Africa, Argentina, Bolivia, Brazil, Chile, Colombia, Israel, Italy, Mexico, Spain, USA and Peru. The importance of Opuntia became evident when research showed that CAM plants could have high productivity in dry regions. Due to their high water use efficiency, their aboveground productivity is much higher than any other arid plant species.

CHAPTER 12

Botanical and agricultural description of grasses

Introduction

Natural grasslands are of interest not only for its livestock farming, but also by other aspects, such as: stop water erosion and/or by wind, maintaining biodiversity, recreation and CO_2 capture. In recent years, increasingly important sectors have become awarded about its importance in relation to the environment and improving the quality of life of the human population. Grasses can be classified by its origin, physiology, potential use by different domestic herbivores or its management, among others. Grasses by origin are divided in cultivated grass and natural and semi-natural. The cultivated: are herbaceous grasses that have been planted and have therefore a floral composition in which introduced species dominate. When the time domain allows for its invasion and spontaneous flora species are considered to have naturalized, e.g., that have become natural, despite its origin.

Characteristics of grasses

Bouteloua curtipendula (Michx.) Torr.

Side-oats grama is a bunchy or sod-forming grass (Figure 12.1). Stems in erect, wiry clumps. Purplish, oat-like spikelets uniformly line one side of the stem, bleaching to a tan color in the fall. The basal foliage often turns shades of purple and red in fall. This is a perennial warm season grass; clump forming. Two varieties are documented: variety curtipendula is shorter and more rhizomatous and ranges from southern Canada to Argentina. Variety caespitosa spreads more by seed than by rhizomes, is more of a bunchgrass, and is restricted mostly to southwestern North America. Its growth begins in early spring. It has two growth forms: 1) low growing (8 to 14 inches) rhizomatous produces few seedheads and reproduces from rhizomes and 2) tall upright (16 to 30 inches) bunch grass produces many seedheads and reproduces from seed. Sideoats grama produces high-quality nutritious forage that is readily eaten by livestock. Wild turkeys eat the seed. This grass responds to proper grazing use and rotation deferred grazing system. It is often a key management species and responds favorably to fertilizer. Good yields of seed can be harvested with a combine. This grass is widely used for seeding depleted ranges and cropland no longer cultivated.

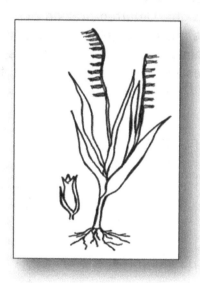

Figure 12.1. *Bouteloua curtipendula* (Michx.) Torr.

Bouteloua trifida (Thruber).

This is a species of grass known by the common name Red Grama. It is native to central and northern Mexico and the Southwestern United

States, where it grows in desert scrub and other dry areas. This is a small perennial grass growing up to about 30 centimeters in maximum height. It sometimes grows a small rhizome. The thready leaves are no more than 5 centimeters long. The inflorescence bears widely spaced spikelets, which are reddish purple in color. Each has three awns at the tip (Figure 12.2).

The distribution of red grama is from West Texas and Oklahoma, Arizona, southern California, and south to central Mexico. This grass is adapted to shallow, gravelly, stony soils that have a poor soil-moisture relationship. Invades deep, loamy soils denuded by overgrazing. Red grama is grazed by all livestock. Forage production is low but quality is good when plants are green. After seedheads appear, nutritive value declines rapidly. This grass helps to protect soil from wind and water erosion if lightly grazed. Growing points and basal leaves are so close to the ground that most livestock cannot harvest more than 50 percent of current year's growth by weight. Deferred grazing during growing season every 2 to 3 years maintains plant vigor and allows plants to mature seed for maintaining a stand. This grass is seldom a key management species except on ranges where it grows in almost pure stands.

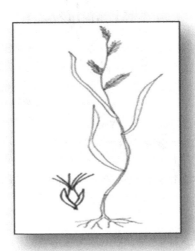

Figure 12.2. *Bouteloua trifida* (Thruber).

Brachiaria fasciculata (Sw.)

Signalgrass is a low growing grass found in open sandy ground of the Edwards Plateau and South Texas Plains and northern Mexico. The culms creep along the ground and root at the nodes. The nodes on the culms are bearded, meaning they have small hairs (Figure 12.3). The leaves are flat. The seed head is approximately 2.5 o 5 cm long. The grass blooms from April to November. Its forage value is poor. The species of this genus are about twenty five, distributed in the tropical and subtropical regions of both hemispheres. Nine species are native to the southern and southwestern United States. The grass is probably the most common species spreading from Florida to Arizona and south at lower elevations to the northern part of South America.

Figure 12.3. *Brachiaria fasciculata* (Sw.)

Cenchrus ciliaris (L.)

Buffel grass is a valuable tropical grass cultivated in Africa, Australia, America and India. The grass is a highly variable, tufted tussock-forming perennial grass. It has a deep, tough rootstock that may go as deep as 2 m. Some varieties are rhizomatous. The culms are erect or decumbent, reaching up to 2 m in length. The leaves are linear blades, green to bluish green, slightly pilose, 3-30 cm long and 4-10 mm wide. The inflorescence is a spike-like panicle, bearing deciduous spikelets, which are surrounded by hairy bristles. The seed is an ovoid caryopsis, 1.4-2 mm long (Figure 12.4).

Cenchrus ciliaris is a very important pasture grass in the tropics. It is cultivated for permanent pastures. The qualities of buffelgrass are numerous. It is easy to establish and provides comparatively high value forage with yields between 2-18 ton DM/ha, without fertilizer, and up to 24 t/ha with the addition of a complete fertilizer. It makes reasonable quality hay when cut in the early flowering stage, yielding up to 2.5 t/ha per cut. Once the seed has been harvested, old grass can give low quality roughage for drought feeding with supplements. The grass is more rarely made into silage, as the moisture content of the grass in semi-arid areas is usually low. This grass is palatable to stock and, once established it can withstand heavy grazing and trampling. In arid areas, it maintains livestock during drought periods. Some strains also grow well during the wet season. It may be a valuable standover feed for winter grazing if supplemented with urea and molasses. There are many cultivars and new genotypes of Cenchrus ciliaris, mostly developed in Australia and North America.

Figure 12.4. *Cenchrus ciliaris* (L.)

Cynodon dactylon (L.) Pers.

Bermuda grass originally came from the savannas of Africa and is the common name for all the East African species of Cynodon. It grows in open areas where there are frequent disturbances such as grazing, flooding, and fire. The blades of Bermuda grass are a grey-green color and are short, usually 2–15 cm long with rough edges (Figure 12.5). The erect stems can grow 1–30 cm tall. The stems are slightly flattened, often tinged purple in color. The seed heads are produced in a cluster of two to six spikes together at the top of the stem, each spike 2–5 cm long. It has a deep root system; in drought situations with penetrable soil, the root system can grow to over 2 m deep, though most of the root mass is less than 60 cm under the surface. The grass creeps along the ground and roots wherever a node touches the ground, forming a dense mat. *Cynodon dactylon* reproduces through seeds, runners, and rhizomes. Growth begins at temperatures above 15 °C with optimum growth between 24 and 37 °C; in winter, the grass becomes dormant and turns brown. Growth is promoted by full sun and retarded by full shade, e.g., close to tree trunks.

Bermuda grass makes good quality hay. As a fine-stemmed leafy species, Bermuda grass cures quickly. It can be tightly packed in bales and maintain good nutritive value during storage. In the USA, Bermuda grass hay is often cubed or pelleted. It should not be cut too late as its nutritive value (protein content) drops with maturity. Six cuts can be taken per year.

Chloris ciliata (Sw.)

This grass is perennial, terrestrial, not aquatic, Stems nodes swollen or brittle, Stems erect or ascending (Figure 12.6). Stems Caespitose, tufted, or clustered. Stems terete, round in cross section, or polygonal. Stem internodes solid or spongy. Stems with inflorescence less than 1 m tall. Stems, culms, or scapes exceeding basal leaves. Leaves mostly cauline. Leaves conspicuously 2-ranked, distichous, Leaves sheathing at base. Leaf sheath mostly open, or loose. Leaf sheath smooth, glabrous. Leaf sheath hairy, hispid or prickly, Leaf sheath or blade keeled, Leaf sheath and blade differentiated. Leaf blades linear. Leaf blades 2-10 mm wide. Leaf blades mostly flat, Leaf blades mostly glabrous. Leaf blades scabrous, roughened, or wrinkled. Ligule a fringe of hairs. Inflorescence terminal. Inflorescence solitary, with 1 spike, fascicle, glomerule, head, or cluster per stem or culm. Inflorescence a panicle with narrowly racemose or spicate branches, Inflorescence a panicle with digitately arranged spicate branches. Inflorescence with 2-10 branches. Inflorescence branches 1-sided. Lower panicle branches whorled. Inflorescence branches paired or digitate at a single node. Flowers bisexual. Spikelets sessile or subsessile. Spikelets laterally compressed, Spikelet less than 3 mm wide, Spikelets with 1 fertile floret, Spikelets with 3-7 florets. Spikelet with 1 fertile floret and 1-2 sterile florets. Spikelets solitary at rachis nodes. Spikelets all alike and fertile. Spikelets bisexual. Spikelets disarticulating above the glumes, glumes persistent. Spikelets disarticulating beneath or between the florets. Rachilla or pedicel glabrous.

Figure 12.6. *Chloris ciliata* (Sw.)

Glumes present, empty bracts. Glumes 2 clearly present. Glumes distinctly unequal. Glumes shorter than adjacent lemma. Glumes keeled or winged. Glumes 1 nerved. Lemma similar in texture to glumes. Lemma 3 nerved. Lemma apex truncate, rounded, or obtuse. Lemma apex dentate, 2-fid. Lemma distinctly awned more than 2-3 mm. Lemma with 1 awn. Lemma awn less than 1 cm long. Lemma awn subapical or dorsal. Lemma awns straight or curved to base. Lemma margins thin, lying flat, Lemma straight. Palea present well developed. Palea shorter than lemma. Palea 2 nerved or 2 keeled. Stamens 3, Styles 2-fid, deeply 2-branched. Stigmas 2, Fruit caryopsis. Caryopsis ellipsoid, longitudinally grooved, hilum long linear. Generally, this grass is distributed in Texas, northern Mexico, Argentina, Uruguay, Cuba and other Caribbean islands. The flowering period runs from March to October. However, this has low nutritional value for grazing ruminants.

Dichanthium annulatum (Forssk.) Stapf in Prain

This grass is tufted perennial to 60 cm (Figure 12.7); the nodes bearded; leaves papillose-pilose at least on the upper surface; first glume of the sessile spikelet not indurate, or slightly indurate. Two to six racemes, sometimes more. Lower glume of sessile spikelet with tubercle-based hairs toward the tip, oblong, obtuse or truncate, keel not winged. Median nerve present, sheaths terete, ligule longish. It differs from *D. caricosum* in having the first glume keeled, not winged, a medial nerve, and large membranous ligule. Ninety-six percent of its roots end within a depth of 1 m. It differs from Bothriochloa pertusa in having no pitting on the glumes and from *Dichanthium sericeum* by the spikelets having a naked appearance due to the hairs being few or almost absent. The spikelets are also very blunt at the top. Roots penetrate to 100 cm in alluvial soil at Varanasi, India, with a yield of 11 275 kg/ha of oven-dried roots.

This grass is a popular pasture grass in many areas. It can be used in fields for grazing livestock, and cut for hay and silage. It is tolerant of varied soil conditions, including soils high in clay and sand, poorly drained soils, and soils that are somewhat alkaline and saline. It forms a turf that can stand up to grazing pressure. It can recover from fire and drought, but it is less tolerant of frost and shade. It does not require fertilizer but it does respond well to a small amount of supplemental nitrogen. It very palatable by cattle. Suitable for silage and hay if cut before flowering. Shows promise for re-seeding degraded grasslands and for rough lawns.

Digitaria insularis (L.).

Digitaria insularis is a tufted perennial with very short, swollen rhizomes (Figure 12.8). The stems reach a height of 80–130 cm and are erect, branched from the lower and middle nodes, swollen bases, with woolly bracts, glabrous internodes and nodes. Sheaths papillose - pilose in their majority, ligule 4–6 mm long, blades linear; 20–50 cm long and 10–20 mm wide. Inflorescence 20–35 cm long; numerous clusters 10–15 cm long; solitary triquetrous rachis of clusters 0.4-0.7 mm wide scabrous; spikelets lanceolate 4.2-4.6 mm long, paired; caudate, densely covered with trichromes up to 6 mm long; brown or whitish, ranging up to 5 mm from the apex of the spikelet; lower glume triangular to ovate to 0.6 mm long; enervate membranous; upper glume 3.5-4.5 mm long; acute, 3-5 nerved, ciliated; inferior lemma as long as spikelet, acuminate, 7-nerved, covered with silky hairs; upper lemma 3.2-3.6 mm long, acuminate, dark brown; anthers 1-1.2 mm long. This species has very low nutritional value for livestock.

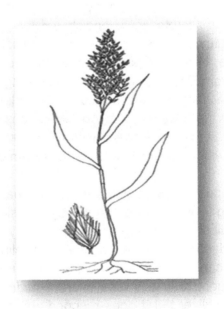

Figure 12.8. *Digitaria insularis* (L.).

Hilaria belangeri (Steud.) Nash.

This grass is native to Mexico and the southwestern United States from Arizona to Texas. This perennial grass forms tufts of stems growing up to about 30 centimeters tall. It forms a sod. The grass has been known to spread four meters in one season. This is the main method of reproduction in the plant because it is often sterile and rarely forms seeds. Growth starts in late spring. Seedheads emerge 30 to 40 days later. Reproduces primarily from stolons. Some stolons are aerial and produce leaves and no roots; others creep along the ground and produce both leaves and roots at nodes, which usually have a ring of hair. Grows mostly in pure stands. Sometimes grows in clusters from axillary buds on basal nodes. Does not tolerate shade. Plants are pale green. This grass is not a vigorous competitor (Figure 12.9).

This grass grows on a wide variety of soils. Grows best on loams to clay loams that have a pH of 6.8 to7.4. It is an important forage for animals in some local regions. Cattle find it very palatable. Wild ungulates such as pronghorn and deer graze upon it. The grass is tolerant of grazing pressure, and even overgrazing. In some areas it is productive early in the season, but most of its productivity occurs after summer rainfall. For maximum production, this grass requires proper grazing use and periodic grazing deferments of 30 to 40 days all year.

Leptochloa filiformis (Lam.).

This grass is a strongly tufted, annual or short-lived perennial grass with glabrous leaves and fibrous roots. Its flowering culms are erect or ascending from a branching base (Figure 12.10). They are occasionally stoloniferous and 0.3-1.2 m tall. They have 3-6 nodes which have glabrous internodes that are smooth, grooved and striate and hollow. The leaf sheaths are keeled, glabrous, smooth, not ciliate, distinctly nerved and usually longer than the associated internodes. The leaf blade is linear, acute, membraneous, green and slightly glaucous. It is 6-32 cm long, 4-9 mm wide, flat or folded, glabrous and rough on the upper surface.The ligules are 1-2 mm long with setaceous hairs on the adaxial surface. They are deeply divided into hairlike segments and erose. The inflorescence forms an open panicle 15-60 cm long, with numerous, slender, flexuous branches. Axis is 10-40 cm long, grooved and scabrous with glabrous axils. The panicle has ascending primary branches which are slender and not winged. They are 2-13 cm long, distinctly grooved and bear spikelets to the base. The pedicels are 0.4-1 mm long and minutely scabrous on the margins.

The spikelets have between 3 and 7 flowers but usually 5-6 flowers are present. These are 2.0-3.5 mm long, 0.8-1.3 mm wide and are often purplish and appressed to the primary branches. The rachilla is filiform and glabrous with similar florets which are perfect, reduced upwards and overlapping. The glumes are shorter than the lemmas. The first glume is usually shorter than the second. It is has a single nerve and is 0.6-1.5 mm long. It is hyaline to membraneous but scabrous on the nerve and lanceolate. The apex is acuminate. The second glume is 0.9-2.4 mm long and 3-nerved. It is keeled and otherwise similar to the lower glume. Disarticulation occurs above the glumes and between the florets. The lemmas are 3-nerved, not deeply cleft; they are 0.8-3-2 mm long and 0.4-0.55 mm wide. Lemmas are membraneous to hyaline with minutely scabrous nerves. They are hairy on the surface and margins and oblong to elliptic in shape with acute to obtuse points. The palea, which is shorter than the lemma, is scabrous on the nerves with appressed hairs. The anthers are minute (about 0.15-0.2 mm long). The caryopsis (grain) is brown, smooth or finely reticulated (rugose). It is 0.5-0.8 mm long, oblong to elliptic and rounded. The caryopsis is dorsally compressed.

This species is grazed readily by all livestock, especially when is green and succulent. During dormant season, it furnishes good quality forage but should be supplemented with a protein concentrate. It grows most often with other grasses. The grass decreases quickly under heavy grazing, but is generally not considered an important forage species. Summer and fall grazing deferments of at least 90 days improve vigor, increase seed production, and provide forage for winter use.

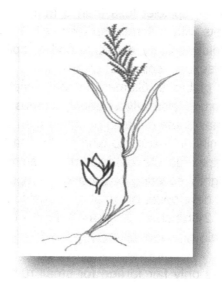

Figure 12.10. *Leptochloa filiformis* (Lam.).

Panicum hallii (Vasey)

This species is perennial (Figure 12.11), terrestrial, not aquatic, stems nodes swollen or brittle, stems erect or ascending, stems geniculate, decumbent, or lax, sometimes rooting at nodes, stems caespitose, tufted, or clustered, stems terete, round in cross section, or polygonal, stems branching above base or distally at nodes, stem nodes bearded or hairy, stem internodes hollow, stems with inflorescence less than 1 m tall, stems, culms, or scapes exceeding basal leaves. Leaves mostly basal, below middle of stem, leaves conspicuously 2-ranked, distichous, leaves sheathing at base, leaf sheath mostly open, or loose, leaf sheath smooth, glabrous, leaf sheath hairy, hispid or prickly, leaf sheath hairy at summit, throat, or collar, leaf sheath and blade differentiated, leaf blades linear, leaf blades 2-10 mm wide, leaf blades mostly flat, leaf blades mostly glabrous, leaf blades more or less hairy, leaf blades scabrous, roughened, or wrinkled, leaf blades glauco us, blue-green, or grey, or with white glands, ligule present, ligule a fringe of hairs. Inflorescence terminal, inflorescence an open panicle, openly paniculate, branches spreading, Inflorescence solitary, with 1 spike, fascicle, glomerule, head, or cluster per stem or culm, inflorescence branches more than 10 to numerous. Flowers bisexual, spikelets pedicellate, spikelets dorsally compressed or terete, spikelet less than 3 mm wide, spikelets with 1 fertile floret, spikelets with 2 florets, spikelet with 1 fertile floret and 1-2 sterile florets, spikelets solitary at rachis nodes, spikelets all alike and fertille, spikelets bisexual, spikelets disarticulating below the glumes. Rachilla or pedicel glabrous. Glumes present, empty bracts, glumes 2 clearly present, glumes distinctly unequal, glumes equal to or longer than adjacent lemma, glume equal to or longer than spikelet, glumes 4-7 nerved. Lemma similar in texture to glumes, lemma 5-7 nerved, lemma glabrous, lemma apex ac ute or acuminate, lemma awnless, lemma margins inrolled, tightly covering palea and caryopsis, Lemma straight. Palea present, well developed, palea about equal to lemma. Stamens 3, styles 2-fid, deeply 2-branched. Stigmas 2. Fruit - caryopsis, Caryopsis ellipsoid, longitudinally grooved, hilum long-linear.

This grass provides only fair forage for small ruminants and wildlife. However, it is a good seed producer, and its seeds can provide a source of food for birds and other wildlife. This grass should not be grazed the first year. After a stand is established, either continuous or rotational grazing can be used. Plants should be allowed to produce seed each year because of the short-lived nature of the plant.

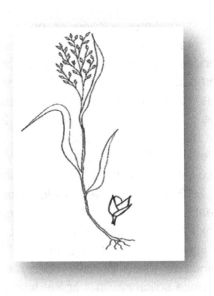

Figure 12.11. *Panicum hallii* (Vasey)

Panicum obtusum (H.B.K.).

It occurs from southern Missouri west to southern Utah, and south to Arizona, New Mexico, Texas, and northern Mexico. This grass is a native, warm season, stoloniferous perennial (Figure 12.12). The height ranges from 30 to 100 cm. The leaf blade is long; narrow, upright, and smooth. The leaf sheath is mostly basal and one half to three fourths as long as the internodes. The stolon is several feet long with long internodes, the nodes swollen and covered with hair. The seedhead has a narrow panicle 2.5 to 10 cm long with the spikelets large, nearly round, and brownish.

Grazing is the primary use of this grass, but it has been cut for hay. It is used to control erosion in waterways and small gullies. Quail and doves eat the seed in fall and early winter. This grass is seldom abundant enough to be a key management species. Generally, it is less palatable than most grasses associated with it. Therefore, it is seldom overgrazed. When used for hay or erosion control, grazing should be delayed until after seed production.

Figure 12.12. *Panicum obtusum* (H.B.K.).

Panicum coloratum (L.)

This plant is a perennial bunchgrass, which usually has rhizomes. It is fine-stemmed and leafy at maturity which culms are erect, 50-120 cm tall, from a knotty base. Leave sheaths glabrous or with papillose based hairs and blades 2.5 mm wide, with scattered papillose based hairs on margins. Panicle is 7 to 20 cm long, spikelets on short pedicels. Spikelets are glabrous, 2.6 to 3.1 mm long with 2 florets which lower floret staminate, with long palea and upper floret fertile, glabrous, shiny, and hard, with acute apex (Figure 12.13).

This grass is used as a pasture grass and to make hay. It produces a large amount of forage for animals. It is drought-tolerant and does well in hot climates. This C4 plant can grow on saline soils and requires an amount of sodium for effective photosynthesis. Different cultivars have varying tolerances of sodium. While it makes a good graze for animals, the grass has occasionally been associated with liver damage and photosensitivity in young ruminants and horses.

Paspalum unispicatum (Scribn. & Merr) Nash

This grass is known for its prominent V-shaped inflorescence consisting of two spike-like racemes containing multiple tiny spikelets, each about 2.8–3.5 mm long. This grass is low-growing and creeping with stolons and stout, scaly rhizomes. The stolons are pressed firmly to the ground and root freely from the internodes, forming a dense sod. The flat, tough-textured leaves are usually hairless, with blades 2–6 mm wide. They are flat, folded, and inrolled, tapering to a fine point. The leaf bases at the terminus of each rhizome usually have a purplish hue. The stems reach 20–75 cm tall. The terminal dual racemes are each attached to the top of a slender stem or with one slightly below the other. There is occasionally a third. The spikelets closely overlap in two rows. They are broad, rounded, smooth and shiny. Inside each spikelet is a tiny flower. The tiny, black, featherlike stigmas and black stamens can be seen dangling at the tips of the flowers (Figure 12.14).

Its distribution generally ranges from southern Texas and Cuba to Venezuela and Argentina. Their flowering periods are during summer and autumn. It is a genus of about four hundred species distributed in all hot regions. Forty-five native species were reported in the southeastern United States. Despite the relatively large number of species, few are especially important as forage grasses in the United States and northern Mexico.

Figure 12.14. *Paspalum unispicatum* (Scribn. & Merr) Nash

Rhynchelytrum repens (Willd.) Hubb.

Perennial, sometimes cespitose. Stem decumbent to erect, 3–10 dm; nodes 4–5. Leaf: sheath 3–9 cm, glabrous; ligule hairs 0.5–1.5 mm; blade 3–20 cm, 3–6 mm wide, upper surface glabrous. Inflorescence 8–17 cm; 1° branches 2.5–6 cm, glabrous to puberulent; spikelet stalk 0.5–5 mm, ± wiry. Spikelet ± 3–5.5 mm, ± 1–2 mm wide, ovate to elliptic; lower glume < 1.5 mm, 0–1-veined, upper glume ± = spikelet, silky hairs white to purplish; lower floret staminate, lemma ± like upper glume, 5-veined, tip minutely lobed, palea ± = lemma; upper floret ± 2/3 length lower floret, lemma firm, ± white, shiny (Figure 12.15).

This grass was introduced to Africa. It is considered invasive in grasslands of the entity. It grows in summer and winter rains. Its forage value is regular. It is relatively unpalatable to livestock.

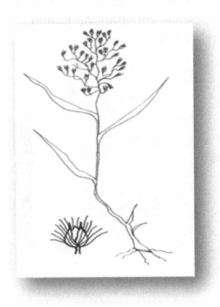

Figure 12.15. *Rhynchelytrum repens* (Willd.) Hubb

Setaria grisebachii (Fourn.).

Annual grass. Culms 30-100 cm; nodes pubescent, hairs appressed. Sheaths with ciliate margins; ligules ciliate; blades to 12(25) cm long, to 10(20) mm wide, flat, hispid on both surfaces (Figure 12.16). Panicles 3-18 cm, loosely spicate, interrupted, often purple; rachises hispid; bristles 1-3, 5-15 mm, flexible, antrorsely scabrous. Spikelets 1.5-2.2 mm. Lower glumes about 1/3 as long as the spikelets, distinctly 3-veined, lateral veins coalescing with the central veins below the apices; upper glumes nearly equaling the upper lemmas, obtuse, 5-veined; lower lemmas equaling the upper lemmas; lower paleas about 1/3 as long as the lower lemmas, narrow; upper lemmas finely and transversely rugose; upper paleas similar to the upper lemmas. 2n = unknown.

Setaria grisebachii is the most widespread and abundant native annual species of *Setaria* in the southwestern United States. It grows in open ground and extends along the central highlands of Mexico to Guatemala, usually at elevations of 750-2500 m. It has low nutritional value for livestock and wildlife.

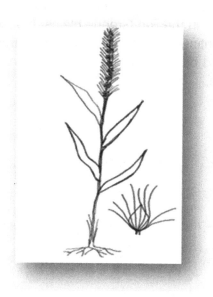

Figure 12.16. *Setaria grisebachii* (Fourn.).

Setaria macrostachya (H.B.K.).

Perennial plant; densely cespitose. Culms 60-120 cm, rarely branched distally, scabrous below the nodes and panicles (Figure 12.17). Sheaths keeled, glabrous, usually with a few white hairs at the throat; ligules 2-4 mm, densely ciliate; blades 15-20 cm long, 7-15 mm wide, flat, adaxial surface scabrous. Panicles 10-30 cm long, 1-2 cm wide, uniformly thick from the base to the apex, dense, rarely lobed basally; rachises scabrous and loosely pilose; bristles usually solitary, 10-20 mm, soft, antrorsely scabrous. Spikelets 2-2.3 mm, subspherical. Lower glumes 1/3-1/2 as long as the spikelets, 3-5-veined; upper glumes about 3/4 as long as the spikelets, 5-7-veined; lower lemmas equaling the upper lemmas, 5-veined; lower paleas nearly equaling the upper paleas in length and width; upper lemmas transversely rugose; upper paleas convex, ovate. 2n = 54.

This grass is grazed by all livestock. It provides high- quality forage when green and succulent but becomes strawy after maturity. If grazed during dormant period, a mineral supplement should be provided. Birds eat the seed. This grass is used in seeding mixtures on some range sites in the Rio Grande plains. It decreases on ranges that are grazed continuously. To maintain vigor and produce a seed crop, grazing should be delayed every 2 to 3 years for 80 to 90 days before seed mature.

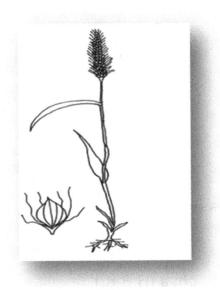

Figure 12.17. *Setaria macrostachya* (H.B.K.).

Tridens eragrostoides (Vasey & Scribn.) Nash.

Plant Caespitose, with knotty, shortly rhizomatous bases (Figure 12.18). Culms 50-100 cm; nodes sometimes sparsely bearded. Sheaths glabrous, scabrous, or sparsely pilose, rounded; ligules 1.2-3 mm, glabrous, membranous, usually lacerate; blades 10-15 cm long, 1.5-5 mm wide, scabrous (occasionally sparsely pilose), apices long attenuate. Panicles 10-30 cm long, to 20 cm wide, open;

branches 5-10(12) cm, lax, ascending to reflexed at maturity, proximal internodes longer than the distal internodes; pedicels (1.5)3-5 mm. Spikelets 3-7 mm, with 5-12 florets. Glumes glabrous, 1-veined, purple; lower glumes 2-2.5 mm; upper glumes 2-3.5 mm; lemmas 2-3.2 mm, veins puberulent to well above midlength, midveins sometimes excurrent, lateral veins rarely reaching the distal margins; paleas 1.5-2 mm, glabrous or scabrous basally, neither enlarged nor bowed-out; anthers 1-1.5 mm. Caryopses 1-1.3 mm. 2n = 40.

Tridens eragrostoides grows in brush grasslands, generally in partial shade. Its range extends from the southern United States into Mexico and Cuba.

Figure 12.18. *Tridens eragrostoides* (Vasey & Scribn.) Nash.

Tridens muticus (Torr.) Wash.

It is a species of grass known by the common name slim tridens. It is native to Mexico and the southwestern quadrant of the United States, where it grows several types of habitat, including plateau and desert, woodlands, sagebrush, plains, and other areas with dry sandy and clay soils (Figure 12.19). It is a perennial grass forming a thick tuft with a knotted base and rhizome. It reaches a maximum height of 50 to 80 centimeters. The panicle has short branches appressed to the others, making the inflorescence narrow. The florets are generally purple in color. This plant uses C4 carbon fixation as its method of energy metabolism.

This species is grazed by cattle and horses. The seed are eaten by rodents and birds. Although this grass makes up 10 to 15 percent of the total production on some sites, it is seldom considered a key manage-ment species because associated grasses are more palatable. It is an increaser on cattle ranges; its abundance indicates fair to poor range condition.

Figure 12.19. *Tridens muticus* (Torr.) Wash.

CHAPTER 13

References

Ackerman-Beetle, A. y Johnson-Gordon, D. 1991. Gramíneas de Sonora. SAGRH COTECOCA. Gobierno del Edo de Sonora, Hermosillo, Son.

Aguado-Santacruz, G.A., Rascón-Cruz, G. Pons-Herández, J.L., Grajeda-Cabrera, O. y García-Moya, E. 2004. Manejo biotecnológico de gramíneas forrajeras. Técnica Pecuaria en México. 42: 261-276.

Akin, D.E., Rigsby, L.L., Lyon, C.E. y Windham, W.R. 1990. Relationship of tissue digestion to textural strength in bermudagrass and alfalfa stems. Crop Sci. 30:990-993.

Akin, D.E., y Chesson, A. 1990. Lignification as the major factor limiting forage feeding value especially in warm conditions. p. 1753-1760. In Proc. XVI int. Grassland Congr., Vol. III. Nice, France, 4-11 October 1989, Association Francaise pour la production Fourragere, Versailles, France.

Albretch, K.A. Wedin, W.F. y Buxton, D.r. 1987. Cell-wall composition and digestibility of alfalfa ítems and leaves. Cerp Sci. 27: 735-741.

Allen, M.S. 1996. Physical constraints on voluntary intake of forages by ruminants. J. Anim. Sci. 74:3063-3075.

Aman, P. 1993. Composition and structure of cell wall polysaccharides in forages. In: H.G. Jung, D.R. Buxton, R.D. Hatfield, and J. Ralph (editores). Forage Cell Wall Structure and Digestibility ASA-CSSA-SSSA, Madison, WI. EUA. p 183.

Aman, P., y Graham, H. 1990. Chemical evaluation of polysaccharides in animal feeds. p. 161-177. In J. Wiseman and D.J.A. Cole (ed.) Feedstuff Evaluation. Butterwarths, London.

Ammerman, C.B., Baker, D.H. y Lewis, A.J. 1995. Bioavailability of Nutrients for Animals: Amino Acids, Minerals and Vitamins. Academic Press, San Diego, EUA. p. 83.

AOAC. 1997. Official Methods of Analysis 17th edición. Association of Agricultural Chemists, Washington, DC.

Armienta-Trejo, G.T. 1995. Perfil mineral del suelo, forraje y tejidos del ganado en agostaderos del Estado de Nuevo León. Tesis de Doctorado. Facultad de Medicina Veterinaria y Zootecnia, Universidad Autónoma de Nuevo León. San Nicolás, N.L., México.

Asplund, J.M. 1994. Principles of protein Nutrition of ruminants. CRC Press. Boca Raton. EUA. pp. 3-28.

Baldwin, R.L., y Allison, M.J. 1983. Rumen Metabolism. J. Anim. Sci. 57 (suplemento 2): 461-477.

Bennetzen, J.L., M. Freeling. 1993. Grasses as a single genetic system: Genome composition, collinearity and compatibility. Trends Genet. 9: 259-261

Barker, D.J., Sullivan, C.Y. y Moser, L.E. 1993 Water deficit effects on osmotic potential, cell wall elasticity, and proline in five grasses. Agron. J. 85:270-275.

Bartnicki-Garcia, S. 1984, Kindoms with walls. pol-8. In W. Dogger and S. Bartnicki-Garcia (ed.) Structure, function, and biosynthesis of plant cell walls. Proe. 7th Ann. Sym. in Botany, Univ. of CA, Riverside, 12-14 Juanary, 1984.

Belsky, A.J. 1992. Effects of trees on nutritional quality of understory gramineous forage in tropical savannas. Tropical Grassl. 26: 12-20

Berlijn J. D., A.E. Bernardón, F.R. Kirchner- S., C.R. Usami-O, E. López-G. 1982. Pastizales Naturales. Editorial Trillas, S.A., México, D.F. pp. 39-40.

Bidlack, J.E., and D.R. Buxton. 1992. Content and deposition rates of cellulose, hemicellulose, and lignin during regrowth of forage grasses and legumes. Can. J. Plant Sci. J. Plant Sci. 72:807-88818.

Blevins, D.G. 1994. Uptake, translocation and function of essential mineral elements in crop plants. In: Boote, K.J. et al. (editors) Physiology and Determiantion of Crop Yied. American Society of Agronomy, Madison, EUA, pp. 259-275.

Blum, A. 1993. Selection for sustained production in water-deficit environments. In D.R. Buxton, R.M. Shibles, R.A. Forsberg, B.L. K.H. Asay, G.M. Paulsen, and R.F. Wilson (ed.) International crop science I. CSSA, Madison, WI.

Blümmel, M. Steingass, H. y Becker, K. 1997. The relationship between in vitro gas production, in vitro microbial biomass yield and 15N incorporation and its implications for the prediction of voluntary feed intake of roughages. British J. of Nutrition. 77: 911-921.

Bowen, E. J. 1981. Hierro: elemento vital para las plantas y animales. Agricultura de las Américas 30:36-41.

Bowen, E. J. y B.A; Krtky. 1983. Microelementos: Causas de deficiencia y toxicidad. Agricultura de las Américas. 36: 8-11.

Briske, D.D. 1991. Development morphology and physiology of grasses. In: Grazing Management. An ecological perspective. Timber Press, Oregon, USA.

Broderick, G.A. 1983. Estimation of protein degradation using in situ and in vitro methods in protein requirements for cattle. F. N. Owens, editor. Symposium, at Oklahoma State Univ., Stillwater, pp. 72-80

Broderick, G.A. 1994. Quantifying forage protein quality. In. Forage quality, evaluation, and utilization. Fahey, C.G. jr. (editor). National Conference on Forage Quality. Univ. of Nebraska-Lincoln. Lincoln, Nebraska, EUA. pp: 200-Brooker R.J., Widmaier E.P., Graham L.E. & Stiling P.D. 2008. Biology. McGraw-Hill. New York.

Buchanan-Smith, J.G. 1990. An investigation into palatability as a factor responsible for the reduced intake of silage by sheep. Animal Production, 50:253-260.

Buxton, D.R. y Fritz, J.O. 1985. Digestibility and chemical composition of cell walls from grass and legume stems. In Proc. 15th. Int. Grassl. Science Council of Japan, Nishi-nasuno, Tochigi-ken, Japan. pp. 1023-1024

Buxton, D.R. y Russell, J.R. 1988. Lignin constituents and cell wall digestibility of grass and legume stems. Crop Sci. 28:553-560.

Buxton, D.R., and M.R. Brasche. 1991. Digestibility of structural carbohydrates in cool-season grass and legume forages. Crop Sci. 31:1338-1345.

Buxton, D.R., Russell, J.R. y Wedin, W.F. 1987. Structural neutral sugars in legume and grass stems in relation to digestibility. Crop. Sci. 27: 1279-1285.

Buxton, D.R., y Casler, M.D. 1993. Environmental and genetic effects on cell-wall composition and digestibility. p. 685-714. In: H.G. Jung, D.R. Buxton, R.D. Hatfield, and J. Ralph (editores) Forage cell wall structure and digestibility. ASA, CSSA, and SSSA, Madison, WI.

Buxton, D.R., y Fales, S.L. 1994. Plant Environment and Quality. En: G.C. Fahey, Jr. editor. Conference on Forage Quality, Evaluation and Utilization. Lincoln, NE. E.U.A.

Casler, M. D.; Talbert, H,; Forney, A.K.; Ehlke, N.J. and Reich, J.M. 1987 Genetic variation for rate of cell wall digestibility and related traits in first cut smooth bromegrass. Crop. Sci. 27.935-939.

Castillo, T. E. Ruiz, G. Febles, A. Barrientos, R. Puente, E. Díaz, G. Bernal. 1993. Aprovechamiento del guaje en Utilización de guaje *Leucaena leucocephala* para la producción de carne bovina en sistemas de bancos de proteína con libre acceso. Revista Cubana de Ciencia Agrícola. 27: 39-44.

Chalupa, W. 1977. Manipulating rumen fermentation. J. Anim. Sci. 46:585-599.

Chapman, G.P. 1996. The biology of Grasses. CAB International. pp 126-140.

Chapman, G.P., Peat, W.E. 1992. An Introduction to the Grasses. CAB Internat., Oxon, UK.

Cheplick, G.P. 1998. Population Biology of Grasses. Cambridge University Press, Cambridge.

Cherney, D.J.R. 2000. Characterization of forages by chemical analysis. In: Forage evaluation in ruminant nutrition. CABI Publishing, Wallingford, UK. pp. 281-300.

Cherney, J.H., Axtell, J.D., Hassen, M.M. y Anliker, K.S. 1988 Forage quality characterization of a chemically induced brown-midrib mutant in pearl millet. Crop Sci. 28: 783-787.

Chesson, A. 1990. Nutritional significance and nutritive value of plant polyssacharides. En: J. Wiseman y D.J.A. Cole (editores). Feedstuff Evaluation. Butterworths, Inc., London RU. pp. 179-195.

Chesson, A. 1993. Mechanistic models of forage. cell wall degradation. p. 347-376. In H. G. Jung, D.R. Buxton, R.D. Hatfield, and J. Ralph (ed.) Forage cell wall structure and digestibility. ASA-CSSA-SSSA, Madison, WI.

Chesson, A. y Fpresberg, C.W. 1988. Polysaccharide degradation by rumen microorganisms. En: P.N. Hobson (editor). The Rumen Microbial Ecosystem. Elsevier Applied Science. Londres, RU. pp. 251-284.

Church, D.C. 1988. The Ruminant Animal Digestive Physiology and Nutrition. Waveland Press, Inc. Illinois, Estados Unidos.

Clark, F.E. y Woodmansee, R.G. 1992. Nutrient cycling. In: Coupland, R.G. (editores) Natural Grasslands: Introduction and Western Hemisphere, Elsevier, Amsterdam, pp. 137-146.

Clark, L.G., W. Zhang, J.F. Wendel. 1995. A phylogeny of the grass family (Poaceae) based on ndhF sequence data. Syst. Bot. 20: 436-460

Cobio-Nagao, C. 2004. Composición química y digestibilidad de la materia seca de pastos nativos colectados en Gral. Terán, N. L. Tesis. Facultad de Ciencias Biológicas, Universidad Autónoma de Nuevo León.

Coleman, G. 1980 Rumen Ciliate Protozoa. Advanced Parasitology. 18:121-173

Conforth, I.S., 1984. Mineral nutrients in pasture species. Proceedings of The New Zealand Society of Animal Production: 44: 135-137.

Corah, L.R. 1990. Ammoniation of low quality forages. Agric. Pract. 11: 35-42.

Crampling, R.C. Processing cereal grain for cattle - a review. Livestock Prod. Sci. 19: 47-60.

Cronjé, P.; E.A. Boomker (2000). Ruminant Physiology: Digestion, Metabolism, Growth, and Reproduction. Wallingford, Oxfordshire, UK: CABI Publishing228.

Cunningham J. 1995. Fisiología Veterinaria Interamericana Mc Graw-Hill. México.

Czarnecka, M., Czarnecki, Z. y Zuromska, M. 1990. Effects of chemical preservation of high-moisture grain on carbohydrates and nitrogen compounds: in vitro experiments. Anim. Feed Sci. and Trech. 34: 343-350.

Da Silva, J.H.S., Johnson, W.L. Burns, J.C. y Andrews, C.E. 1987. Growth and environment effects on anatomy and quality of temperate and subtropical forage grasses. Crop Sci. 27:1266-1273.

Datnoff LE, Rutherford BA. 2004. Effects of silicon on leaf spot and melting out in bermudagrass. Golf Course Manage. 5(1): 89-92.

Datnoff L, Brecht M, Kucharek T, Trenholm L, Nagata R, Snyder G, Unruh B, Cisar J. 2005. Influence of silicon (Si) on controlling gray leaf spot and more in St. Augustinegrass in Florida. TPI Turf News, 3(2): 30-32.

Delmer, D.P., y Store, B.A. 1988. Biosynthesis of plant cell walls. p. 373-420. In J. Preiss (ed.) The biochemistry of plants, vol. 14 Academic Press, New York.

Dittberner, P.L., Olson, M.L. 1983. The plant information network (PIN) data base: Colorado, Montana, North Dakota, Utah, and Wyoming. FWS/OBS-83/86.

Dijkstra, J.; J.M. Forbes, J. France (2005). Quantitative Aspects of Ruminant Digestion and Metabolism, 2nd edition. Wallingford, Oxfordshire, UK: CABI Publishing. p. 736 pages.

Dijkstra J. 2005: Quantitative Aspects of Ruminant Digestion and Metabolism (2nd Edition). CABI Publishing. Wallingford.

Dehority, B.A. 1993. Microbial ecology of cell wall fermentation. p. 425-453. In H. G. Jung, D.R. Buxton, R.D. Hatfield, and J. Ralph (ed.) Forage cell wall structure and digestibility. ASA-CSSA-SSSA, Madison, WI.Dougherty, C.T., Bradley, N.W., Cornelius, P.L. y Lauriault, L.M. 1989. Short term fasts and the ingestivo behavior of grazing cattle. Grass and Forage Science, 40: 69-77.

Duncan, A.J. y Milne, J.A. 1993. Effects of oral administration of brassica secondary metabolitos, allycyanide, allyl isothicynate and dimethyl disulphide, on the voluntary intake and metabolism of sheep. British J. of Nutrition, 70: 631-645.

Duncan P., Foose T.J., Gordon I.J., Gakahu C.G. & Lloyd M. 1990: Comparative nutrient extraction from forages by grazing bovids and equids: a test of the nutritional model of equid/bovid competition and coexistence. Oecologia 84: 411-418.

Durand, M. and R. Kawashima. 1980. Influence of mineral in rumen microbial digestion. In: Digestive Physiology and Metabolism in Ruminants. Avi. Publishing Company, Inc. Westport, Connecticut. p. 375.

Durand, M. y Komisarczuk, S., 1988. Influence of major mineral son rumen microbiota. J. of Nutrition 151:413-426.

Emmans, G.C. 1997. A method to predict food intake of domestic animal from birth to maturity as a function of time. J. of Theoretical Biology 186: 189-199.

Engels, F.M. 1989. Some properties of cell wall layers determining ruminant digestion. In: A. Chesson and E.R. Orskov (editores). Physico-Chemical Characterization of Plant Residues for Industrial and Feed Use. Applied Science Publishers, London. p. 80.

Epstein E. The anomaly of silicon in plant biology. Proc. Natl. Acad. Sci. USA. 1994;91(1):11–17.

FAO. 2009. Grassland Index. Disponible en línea en

Farrar, J.F. 1988. Temperature and the partitioning and translocation of carbon. p. 203-235. In S.P. Long and F.I. Woodward (ed.) Plants and temperature. Soc. Exp. Biol., Cambridge, U.K.

Fick, G.W., Holt, D.A. y Lugg, D.G. 1988. Environmental physiology and crop growth. p. 163-1194. In A.A. Hanson, D.K, Barnes, and R.R. Hill, Jr. (ed.) Alfalfa improvement. Agronomy Monogr. 29 ASA, CSSA, and SSSA, Madison, WI.

Follet, R.F. y Wilkinson, S.R. 1995. Nutrient management forages. In: Barnes, R.F., Miller, D.A. y Nelson, C.J. (editores) Forages, 5a edición, vol. II. Iowa State Univ. Press, Ames, pp. 55-82.

Fontenot, J.P. y Church, D.C., 1979. The macto (major) minerals. In: Church, D.C. (editor). Digestive Physiology and Nutrition of Ruminants, Vol. 2. Nutrition, 2a edición. O & B Books, Corvalis, Oregon, pp. 56-99.

Forbes, J.M. 1995. Voluntary Food Intake and Diet Selectioon in Farm Animals. CAB International, Wallingford, Reino Unido.

Forbes, J.M. 1996. Integration of regulatory signals controlling forage intake of ruminants. J. Anim. Sci. 74: 3029-3035.

Foroughbackhch, R. Ramírez, R.G., Hauad, L., Alba-Avila, J., García-Castillo, C.G. y Espinosa-Vázquez, M. 2001. Dry matter, crude protein and cell wall digestion of total plant, leaves and stems in llano buffelgrass (Cenchrus ciliaris). J. Applied Animal Research. 20:181-188.

Forsberg, C.W., Forano, E. y Chesson, A. 2000. Microbial Adherence to the Plant Cell Wall and Enzymatic Hydrolisis. En: P.B. Cronjé (editor). Ruminant Physiology, Digestion, Metabolism, Growth and Reproduction. CABI, Publishing, Wallingford, Reino Unido. pp. 79-98.

Foth, Dh.H. 1985. Fundamentos de la ciencia del suelo. 3ª. Ed. Compañía Editorial Continental. México. Pp. 313-314.

Ganskoop, D., Bohnert, R. 2001. Nutritional dynamics of 7 northern Great Basin grasses. J. Range Manage., 54:640-647.

García-Dessommes, G.J., Ramírez-Lozano R.G., Foroughbakhch, R., Morales-Rodríguez, R. y García-Díaz G. 2003. Valor nutricional y digestión ruminal de cinco líneas apomíticas del pasto buffel (Cenchrus ciliaris L.). Técnica Pecuaria en México, 41: 209-218.

García-Dessommes, G.J., Ramírez-Lozano, R.G., Foroughbakhch, R., Morales Rodríguez, R. y García-Díaz, G. 2003. Ruminal digestion and chemical composition of new genotypes of buffelgrass (Cenchrus ciliaris L.). Interciencia, 28: 220-224.

García-Dessommes, G.J., Ramírez-Lozano, R.G., Foroughbakhch, R., Morales Rodríguez, R. y García-Díaz, G. 2003. Ruminal digestion and chemical composition of new genotypes of buffelgrass (Cenchrus ciliaris L.). Interciencia, 28: 220-224.

Gissel-Nielsen, G., Grupa, U.C., Lamand, M. y Westermarck, T. 1984. Selenium in soils and plants and its importante in livestock and human nutrition, Advances in Agronomy. 37: 397-460.

Gould, F. W. 1975. The grasses of Texas. Texas A & M University Press.

Gould, F.W. y R.B. Shaw. 1992. Gramíneas Clasificación Sistemática. AGT Editor, S.A. México, D.F. pp. 45-332

Greene, L.W., Pinchak, W.E. y Heitschmidt, R.K. 1987. Seasonal dynamics of mineral in forages at the Texas Experimental Ranch. J. Range manage. 40:502-506.

Halim, R.A., Buxton, D.R. Hattendorf, M.J. y Carlson, R.E. 1989. Water-stress effects on alfalfa forage quality after adjustment for maturity differences. Agron. J. 81:189.

Harris, P.J., Hartley, R.D. y Lowry, K.H. 1980. Phenolic constituents of mesophyll and non-mesophyll cell walls from leaf laminae of Lolium perenne. J. Sci. Food Agric. 31:959-962.

Hatfield R.D., Ralph, J., Grabber, J. y Jung, H.J. 1993. Structural characterization of isolated corn lignins. p. A-319. In Abstr. Keystone Symposia, the extracellular matrix of plants: Molecular, cellular and developmental biology. Santa Fe, NM, 9-15 Jan., 1993.

Hatfield, R.D., Jung, H.G Ralph, J.. Buxton, D.R y Weimer, P.J. 1994. A comparison of the insoluble residues produced by the Klason lignin and acid detergent lignin procedures. J. Sci. Food Agric. 65:51-57.

Hattersley, P.W., Watson, L. 1992. Diversification of phostosynthesis. pp. 58-116. In: Chapman, G.P. editor. Grass Evolution and Domestication. Cambridge University Press. p. 390.

Hendrickson, J.R., Briske, D.D. 1997. Axillary bud banks of two semiarid perennial grasses: occurrence, longevity and contribution to population persistence. Oecologia. 110: 584-591.

Henning, J.C., Dougherty, C.T., O'Leary, J., Collins, M. 1990. Urea for preservation of moist hay. J. Anim. Feed Sci. Tech. 31: 193-199.

Hespell, R.B. 1988. Microbial digestion of hemicelluloses in the rumen. Microbiol. Sci. 5: 362-365.

Hides, D.H., Lovatt, J.A. y Hayward, M.V. 1983. Influence of stage of maturity on the nutritive value of Italian ryegrasses. Grasses Forage Sci. 38: 33-38.

Hobson, P.N.; C.S. Stewart (1997). The Rumen Microbial Ecosystem, 2nd edition. New York: Springer.

Hodgson, J. 1981. The influence of variation in the surface characterisitcs of the sward upon the short-term rate of herbage intake by calves and lambs. Grass Forage Sci. 36:49-57.

Hodgson, J., Fournes, T.D.A., Armstrong, R.H., Beattie, M.M. y Hunter, E.A. 1991. Comparative studies of the ingestivo behavior and herbage intake of sheep and cattle grazing indigenous hill plant communities. J. of Applied Ecology 28: 205-227.

Hofmann, R.R. 1988. Anatomy of the Gastro-Intestinal Tract. En: The ruminant animal, Digestive Physiology and Nutrition. D.C. Church (editor). Waveland Press, Inc. Illinois, Estados Unidos. pp. 14-43.

http://www.fao.org/ag/AGP/AGPC/doc/GBASE/mainmenu.htm

Huber, J.T. y Kung, L. Jr. 1981. Protein and nonprotein nitrogen utilization in dairy cattle. J. Dairy Sci. 64: 1170-1177.

Hungate, R.E. 1966. The Rumen and its Microbes. Academic press, New York.

Hunt, C.W., Paterson, J.A. y Wiliams, J.E. Intake and digestibility of alfalfa-tall fecue combination diets fed to lambs. J. Anim. Sci. 60:301-306.

Huss, D.L., Aguirre, E.L. 1976. Fundamentos de Manejo de Pastizales. Instituto Tecnológico y de Estudios Superiores de Monterrey. pp. 21-45.

Huston, J.E., Rector, B.S., Merril, L.B., Engdahal, B.S. 1981. Nutritional value of range plants in the Edwards Plateau region of Texas. Report B-1375. College Station, TX: Texas A& M University System, Texas Agricultural Experimental Station., p. 16.

Iiyama, K., Lam, T.B.T. y Stone, B.A. 1990. Phenolic acids bridges between polysaccharides and lignin in wheat internodes. Phytochemistry (Oxf.) 29:733-740.

Illus, A.W. 1998. Advances and retreats in specifying the constrains on intake in grazing ruminants. En: Buchanan-Smith, J.G., Bailey, L. y McCaughey, P. (editores) Proceedings of XVIII Internacional Grassland congreso, Vol. III. Association Management Centre, Calgary, Canadá, pp. 39-44.

Illus, A.W. y Gordon, I.J. 1991. Prediciton of intake and digestion in rumiants by a model of rumen kynetics integrating animal size and plant characteristics. J. Agricultural Science, 116: 145-157.

Inanaga S, Higuchi Y, Chishalci N. 2002. Effect of silicon application on reproductive growth of rice plant. Soil Sci Plant Nutr, 48(3): 341-345. Fales, S.L. 1986. Effects of temperature on fiber concentration, composition, and vitro digestion kinetics of tall fescue. Agron. J. 78:963-966.

Ingvartesen, K.L., Anderson, H.R. y Foldager, J. 1992. Effect of sex and pregnancy on feed intake capacity of growing cattle. Acta Agriculturae Scandinavica, Section A, Animal Science, 42: 40-46.

Judd, W.S., Campbell, C. S. Kellogg, E. A. Stevens, P.F. Donoghue, M.J. 2002. Poaceae. Plant systematics: a phylogenetic approach. (Sinauer Axxoc edición). USA. pp. 287–292.

Jung, H.G. y Allen, M.S 1995. A review of lignin in forages. J Anim. Sci. 73:2774-2790.

Jung, H.G., Smith, R.R. y Endres, C.S. 1994. Cell wall composition and degradability of stem tissue from lucerne divergently selected for lignin and in vitro dry matter disappearance. Grass Forage Sci. 49:295.

Jung, H.G., y Russelle, M.P. 1991. Light source and nutrient regime effects on fiber composition and digestibility of forages. Crop Sci. 31:1065.

Kawas-Garza, J.J. 1996. Determinación del perfil mineral de especiees forrajeras de cuatro zonas geográficas del litoral del golfo de México. Tesis de Doctorado. Facultad de Medicina Veterinaria y Zootecnia, Universidad Autónoma de Nuevo León. San Nicolás, N.L., México.

Kennedy, P.M., Christopherson, R.J. y Milligan, L.P. 1982. Effects of cold exposure on feed protein degradation, microbial protein synthesis and transfer of plasma urea to the rumen of sheep. Br. J. Nutr. 47:521.

Kephart, K.D., Buxton, D.R. y Taylor, S.E. 1992. Growth of C3 and C4 perennial grasses in reduced irradiance. Crop Sci. 32:1033-1038.

Kephart, K.D., y Buxton, D.R. 1993. Forage quality responses of C3 and C4 perennial grasses to shade. Crop Sci. 33:831-837.

Kerley, M.S., G.C. Fahey, Jr., J.M. Gould, and E.L. Iannotti. 1988. Effects of lignification, cellulose crystallinity and enzyme accessible space on the digestibility of plant cell wall carbohydrates by the ruminant. Food Microstructure 7: 59-65.

Klopfenstein, T.J. y Owen, F.G. 1981. Value and potential use of crop residues and by-products in dairy rations. J. Dairy Sci. 64: 1250-1257.

Kowalenko, C.G. 1993. Extraction of available sulfur. In: Carter, M.R. (editor). Soil Sampling and Methods of Analysis. Lewis Publishers, Boca Raton, pp. 65-74.

Lam, T.B.T., Iiayama, K. y Stone, B.A. 1990. Primary and secondary walls of grasses and other forage plants: Taxonomic and structural considerations. En D.E. Akin, L.G. Ljungdhal, J.R. Wilson y P.J. Harris (editores). Microbial and Plant Opportunities to Improve Hemicellulose Utilization by Ruminants. Elsevier Sci. Publishing Co., Inc., NY., Estados Unidos. pp. 43-69.

Langelands, P.J., 1987. Assessing the nutrient status of herbivores. In: Hacker, J.B. and Ternouth, J.H. (editores). The Nutrition of Herbivores. Academic Press, Sydney. pp. 363-390.

Langer, P. y Snipes, R.L. 1991. Adaptations of Gut Structure to Function in Herbivores. En: T. Tsuda, Y. Sasaki y R. Kawashima (editors). Physiological Aspects of Digestion and Metabolism in Ruminants. Academic Press, Inc. San Diego, Estados Unidos. pp. 349-384.

Lawler, D.W. 1993. Photosynthesis: Molecular, Physiological and Environmental Processes, 2nd edn. Logman. p. 318.

Lebgue, T. A. Valerio. 1991. Gramíneas de Chihuahua. Manual de Identificación. Colección Textos Universitarios, Universidad Autónoma de Chihuahua, México. pp. 15-40.

Leibowitz, S.F. y Stanley, B.G. 1986. Neurochemical controls of appetite. En: Feeding Behavior and Humeral Controls. R.C. Ritter, S. Ritter y C.P. Barnes (editores). Academic Press, San Diego, California, Estados Unidos, pp. 191-234.

Leng, R.A. 1990. Factors affecting the utilization of poor quality forages by ruminants particularly under tropical conditions. Nutrition Research Reviews 3: 277-303.

Levitt, J. 1980. Response of plants to environmental stresses. Vol. 11. 2nd ed. Academic Press, New York, NY.

Liang Y. 1999. Effect of silicon on enzyme activity, and sodium, potassium and calcium concentration in barley under salt stress. Plant and Soil. 209(2): 217-224.

Mackie, R.I., Aminov, R.I., White, B.A. y McSwenney, C.S. 2000. Molecular Ecology and Diversity in Gut Microbial Ecosystems. En: P.B. Cronjé (editor). Ruminant Physiology, Digestion, Metabolism, Growth and Reproduction. CABI, Publishing, Wallingford, Reino Unido. pp. 61-78.

MacPherson, A. 2000. Trace-mineral status of forages, En: Forage Evaluation in Ruminant Nutrition. D.I. Givens, E. Owen, R.F.E. Axford y H.M. Omed (editores). CAB International. Wallinford, Reino Unido. pp. 345-372.

Mahadevan, S., J. D. Erfle, and F. D. Saver. 1980. Degradation of soluble and insoluble proteins by Bacteriodes amylophylus protease and by rumen microorganisms. J. Anim. Sci. 50:723-730.

McLeay, L.M., B.L. Smith & S.C. Munday-Finch. 1999. Tremorgenic mycotoxins paxilline, penitrem and lolitrem B, the non-tremorgenic

31-epilolitrem B and electromyographic activity of the reticulum and rumen of sheep. Research in Veterinary Science 66: 119-127.

Marschner, H. 1993. Zn uptake from soils. In: Robson, A.D. (editores). Zinc in Soil and Plants. Kluwer Academic Publishers, Dordrecht, pp. 59-77.

Marschner, H. 1995. Mineral Nutrition of Higher Plants, 2a edición, Academic Press, Londres R.U. p. 889.

Marten, G.C., Buxton, D.R. y Barnes, R.F. 1988. Feeding value (forage quality). p. 463-491. In A.A. Hansen, D.K. Barnes, and R.R. Hill, Jr. (ed.) Alfalfa and alfalfa improvement. Agronomy Monogr. 29. ASA, CSSA, and SSSA, Madison, WI. EUA.

Martín-Rivera, M., Ibarra-Flores, F. 1989. Manejo de Pastizales. En: Veinte años de investigación pecuaria en el CIPES. Proyecto No. P89017, Universidad de Sonora, México.

McAllan, A.B., y Smith, R.H. 1983. Estimation of flows of organic matter and nitrogen components in postruminal digesta and effects of level of dietary intake and physical form of protein supplement on such estimates. Br. J. Nutr. 49:119-125.

McBride, M.B. 1994. Environmental Chemistry of Soils. Oxford University Press, N.Y. p. 406.

McDonald, P. 1981. The Biochemistry of Silage. John Wiley & Sons, NY. EUA. p. 167.

McDonald, P., Edwards, R.A. y Greenhalgh, J.F.D. 1995. Animal Nutrition. 5a edición. Logman Scientific & Technical. pp: 132-157.

McDowell, L.R. 2003. Minerals in Animal and Human Nutrition, 2a edición. Elsevier Science, Amsterdam, Holanda. pp. 35-37.

Merchen N.R. y Bourquin, L.D. 1994. Processes of digestion and factors influencing digestion of forage based diets by ruminants. En G.C. Fahey, Jr. (editor). Forage Quality, Evaluation, and Utilization. University of Nebraska, Lincoln, NE. Estados Unidos. Pp. 564-612.

Merchen, N.R. 1988. Digestion, Absorption and Excretion in Ruminants. En: The ruminant animal, Digestive Physiology and Nutrition. D.C. Church (editor). Waveland Press, Inc. Illinois, Estados Unidos. pp. 173-201.

Mertens, D.R. 1993. Kinetics of cell wall digestion and passage in rumiants. p. 535-570. In H. G. Jung, D.R. Buxton, R.D. Hatfield, and J. Ralph (ed.). Forage cell wall structure and digestibility. ASA-CSSA-SSSA, Madison, WI.

Millard, P. Gordon A.H., Richardson, A.J. y Chesson, A. 1987. Reduced ruminal degradation of rygrass caused by J. of Sci. of Food and Agricultural 40: 305-314.

Miller, E.R. 1985. Mineral x disease interactions. J. Anim. Sci. 60: 1500-1507.

Milne, L., Milne, M. 1967. Living Plants of the World. Chaticleer Press, N.Y.

Minson, D.J, 1992, Composición química y valor nutritivo de gramíneas tropicales. En: Gramíneas Tropicales. P.J. Skerman y F. Riveros (editores). Colección FAO: Producción y Protección Vegetal.

Minson, D.J. 1990. Forage in Ruminant Nutrition. Academia Press, San Diego, California, Estados Unidos.

Minson, D.J. y Willson, J.R. 1994. Prediciton of intake as an element of forage quality. En: Fahey, G.C. Jr. (editor). Forage Quality, Evaluation, and Utilization, pp. 533-563

Moe, P.W. 1981. Energy metabolism of dairy cattle. J.Dairy Sci. 64:1120-1139.

Monties, B. 1991. Plant cell as fibrous lignocellulosic composites: Relations with lignin structure and function. Anim. Feed Sci. Technol. 32:159-175.

Moore, H.J. y Jung, H.G. 2001. Lignin and fiber digestion. J. Range Manage. 54: 420-430.

Morales-Rodríguez, R. 2003. Producción de materia seca y digestibilidad in situ del forraje de 84 genotipos del pasto buffel (Cenchrus ciliaris). Tesis de maestría. Facultad de Ciencias Biológicas, Universidad Autónoma de Nuevo León.

Morales-Rodríguez, R. 2003. Producción de materia seca y digestibilidad in situ del forraje de 84 genotipos del pasto buffel (Cenchrus ciliaris). Tesis de maestría. Facultad de Ciencias Biológicas, Universidad Autónoma de Nuevo León.

Morales-Rodríguez, R., Ramírez, R.G., García-Dessommes, G.J. y González-Rodríguez, H. 2005. Nutrient Content and In Situ Disappearance in 78 Genotypes of Buffelgrass (Cenchrus ciliaris L.). J. Appl. Anim. Res. (En prensa).

Murphy, M. 1989. The influence on non-structural carbohydrates on rumen microbes and rumen metabolism in milk producing cows. Ph.D. thesis, Swedish University of Agricultural Sciences, Uppsala.

Nelson, C.J., Moser, L.E. 1994. Plant factors affecting forage quality. En: Forage Quality, Evaluation and Utilization. National. Fahey C.G. Jr (ed.) Conference on Forage Quality. Univ. of Nebraska. Lincoln Nebraska, EUA. P. 123.

Newman, B.D., Wilkox, B.P., Archer, R.R., Breshears, D.D., Dahm, C.J., Duffy, N.G., McDowell, F.M., Phillips, B.R., Scanlon, E.R., Vivoni. C.N. 2006. Ecohydrology of water–limited environments: A scientific vision. Water Resources Research.

Newman, J.A., Parsons, A.J. y Penning, P.D. 1994. A note on the behavioral strategies used by grazing animals to alter their intake rate. Grass and forage Science, 49: 502-505.

Nobel, P.S. 1988. Principles underlying the prediction of temperature in plants with special reference to desert succulents. p. 1-23. In S.P. Long and F.I. Woodward (ed.). Plants and temperature. Soc. Exp. Biol., Cambridge, U.K.

Nocek, J. E., and S. Tamminga. 1991. Site of digestion of starch in the gastrointestinal tract of dairy cows and its effect on mild yield and composition. J. Dairy Sci. 74:3598-3629.

NRC, 1989. National Research Council. Nutrient Requirements of Dairy Cattle, 6a edition. National Academic Press, Washington, DC. p. 157.

NRC, 2000. National Research Council. Nutrient Requirements of Beef Cattle, 7a edition. National Academic Press, Washington, DC. p. 175.

O'Neill, M.A. y York, W.S. 2003. The composition and structure of plant primary cell walls. En: Rose, J. K. C. The Plant Cell Wall. Oxford: Blackwell. pp. 1–54.

Orpin, C. 1983. The role of ciliate protozoa and fungi in the rumen digestion of plant cell walls. Anim. Feed Sci. Technol. 10:121-143

Ørskov, E.R. 2000. The in situ technique for the estimation of forage degradability in ruminants. En: D.I. Givens, E. Owen, R.F.E. Axford, y H.M. Omed (editores). Forage Evaluation in Ruminants, CAB International, Wallinford RU. pp. 175-188.

Ørskov, E.R. y Kay, R.N.B. 1987. Non-microbial digestion of forages by herbivores. En: J.B. Hacker y J.H. Ternouth (editors). The Nutrition of Herbivores. Academic Press, Inc, Orlando Fl. Estados Unidos. pp. 267-280.

Osafo, E.L.K., Owen, E. Said, A.N., Gill, M y Sherington, J. 1997. Effects of the ammont offered and chopping on intake and selection of soghum stover by Ethiopian sheep and cattle. Animal Science, 65: 55-62.

Owens, F.N. y Berger, W.G. 1983. Nitrogen metabolism of ruminant animals: Historical perspective, current understanding and future implications. J. Anim. Sci. 57:498-518.

Parson, A.J. and Penning, P.D. 1988. The effect of the duration of regrowth on photosynthesis, leaf death and the average rate of growth in a rotationally grazed sward. Grass Forage Science, 43:15-27.

Paul, E.A. y Clark, F.E. 1996. Soil Microbiology and Biochemistry, 2ª edición. Academic Press, San Diego.

Pollak, C.J. 1990. The response of plants to temperature change. J. Agric. Sci., Camb. 115:1-5.

Poppi, D. P., Minson, D. J. y Ternouth, J. H. 1981. Studies of cattle and sheep eating leaf and stem fractions of grasses. I. The voluntary intake, digestibility, and retention time in the reticulo-rumen. Aust. J. Agric. Res. 32: 99-108.

Provenza, F.D. 1995a Post ingestive feedback as an elementary determinant of preference and intake of rumiants. J. Range Manage. 48: 2-17.

Provenza, F.D. 1995b. Role of learning in food preferences of ruminants. Greenhalgh and Reid revisited. In: von Engelhardt, W., Leonhard-Marek, S. Breves, G. y Giesecke, D. (editors) Ruminant Physiology: Digestion, Metabolism, Growth and Reproduction Proceedings of the Eight International Symposium on Ruminant Physiology. Ferdinand Enke Verlag, Stuttgardt, Alemania, pp. 233-247.

Puoli, J.R., Jung, G.A. y Reid, R.L. 1991. Effectos of nitrogen and sulfur on digestión and nutritive quality of warm-season grass hays for cattle and sheep. J. Anim. Sci. 69:843-852.

Ralph, J., Hatfield, R.D., Quideau, S., Helm, R.F., Grabber, J.H. y Jung, H.G. 1994. Pathway of p-coumaric acid incorporation into maize lignin as revealed by NMR. J. Am. Chem. Soc. 116:9448-9452.

Ralphs, M.H. y Cheney, C.D. 1993. Influence of cattle age, lithium chloride dose level and food type in the retention of food aversions. J. Anim. Sci. 71: 373-379.

Ramírez R.G., González-Rodríguez, H. y García-Dessommes, G. 2002. Chemical composition and rumen digestion of forage from kleingrass (Panicum coloratum). Interciencia, 27:705-709.

Ramírez R.G., Haenlein, G.F.W., García-Castillo, C.G. y Núñez-González M.A. 2004. Protein, lignin and mineral contents and in situ dry matter digestibility of native Mexican grasses consumed by range goats. Small Rumin. Res. 52: 261-269.

Ramírez R.G., Haenlein, G.F.W., García-Castillo, C.G. y Núñez-González M.A. 2004. Protein, lignin and mineral contents and in situ dry matter digestibility of native Mexican grasses consumed by range goats. Small Rumin. Res. 52: 261-269.

Ramírez R.G., Haenlein, G.F.W., García-Castillo, C.G. y Núñez-González M.A. 2004. Protein, lignin and mineral contents and in situ dry matter digestibility of native Mexican grasses consumed by range goats. Small Rumin. Res. 52: 261-269.

Ramírez, R.G. y González-Rodríguez, H. García-Dessommes, G. and Morales-Rodríguez, R. 2005. Seasonal Trends in the Chemical Composition and Digestion of Dichanthium annulatum (Forssk.) Stapf. J. Appl. Anim. Res. (En prensa).

Ramírez, R.G. y González-Rodríguez, H. García-Dessommes, G. and Morales-Rodríguez, R. 2005. Seasonal Trends in the Chemical Composition and Digestion of Dichanthium annulatum (Forssk.) Stapf. J. Appl. Anim. Res. 25: 12-20.

Ramírez, R.G. y González-Rodríguez, H. García-Dessommes, G. and Morales-Rodríguez, R. 2005. Seasonal Trends in the Chemical Composition and Digestion of Dichanthium annulatum (Forssk.) Stapf. J. Appl. Anim. Res. 28: 35-40.

Ramírez, R.G., Foroughbackhch, R., Háuad, L. Alba-Avila, J. García-Castillo, C.G. y Espinosa-Vázquez, M. 2001. Seasonal variation of in situ digestibility of dry matter, crude protein and cell wall of total plant, leaves and stems of Nueces buffelgrass (Cenchrus ciliaris). J. Appl. Anim. Res., 20:38-47.

Ramírez, R.G., Foroughbackhch, R., Háuad, L. Alba-Avila, J. García-Castillo, C.G. y Espinosa-Vázquez, M. 2001. Seasonal variation of in situ digestibility of dry matter, crude protein and cell wall of total plant, leaves and stems of Nueces buffelgrass (Cenchrus ciliaris). J. Appl. Anim. Res., 20:38-47.

Ramírez, R.G., García-Dessommes, G.J. y González Rodríguez, H. 2003. Valor nutritivo y digestión ruminal del zacate buffel común (Cenchrus ciliaris L.), Pastos y Forrajes 26: 149-158.

Ramírez, R.G., González-Rodríguez, H. y García-Dessommes, G.J. 2003. Nutrient digestion of common bermudagrass (Cynodon dactylon L.) Pers. growing in northeastern Mexico, J. Appl. Anim. Res. 23: 93-102.

Ramírez-Lozano, R.G. 2003. Nutrición de Rumiantes: Sistemas Extensivos. Trillas, México, D.F. p. 212.

Reid, L.M., Jung, G.A. y Allison, D.W. 1988. Nutritive quality of warm season grasses in the Northeast, W.V. Univ. Agric. Forestry Exp. Stn. Bull., p. 699.

Reid, R.L. 1994. Nitrogen compounds of forages and feedstuffs. En: Principles of Protein Nutrition of Ruminants. Asplum, J.M. (editor). CTS Press. Boca Ratón, EUA. pp. 43-64.

Reid, R.L., Jung, G.A. y Thayne, W.A. 1988. Relatioships between nutritive quality and fiber components of cool season and warm season forages: a retrospective study. J. of Anim. Sci. 66: 1275-1291.

Ritter, R.C., Ritter, S. y Barnes, C.P. (editores) 1986. Feeding Behavior and Humeral Controls. Academic Press, San Diego, California, Estados Unidos.

Robbins, C.T. 2001. Wildlife Feeding and Nutrition. 2a edición. Academic Press. San Diego, EUA. p. 248.

Rodrigues FA, Vale FX, Datnoff LE, Prabhu AS. 2003. Effect of rice growth stages and silicon on sheath blight development. Phytopathology. 93(5): 256–261.

Rojas, M. y Rovalo, M. 1986. Fisiología Vegetal. Editorial. McGrawHill. México. p. 115.

Romheld, V. y Marschner, H. 1991. Functions of micronutrients in plants. In: Mortvendt, J.J. (editor) Micronutrients in Agriculture, 2a edición. Soil Science Society of America, Madison, pp. 297-328.

Romney, D.L. y Gill, M. 2000. Intake of forages. En: Forage Evaluation in Ruminant Nutrition. D.I. Givens, E. Owen, R.F.E. Oxford y H.M. Omed (editores). CAB International, Wallingfor, Reino Unido, pp. 43-62.

Romney, D.L., Blunn, V. y Leaver, J.D. 1997. The responses of lactating dairy cows to diets based on grass silage of high or low DM and supplemented with fast and slowly fermentable energy sources. En: Proceedings of the British Society of Animal Science, Edinburgo, Escocia, p. 84.

Rosenberg, N.J., B.L. Blad, and S.B. Verma. 1983. Microclimate: The biological environment. 2nd ed. Wiley Interscience, New York, NY.

Salisbury, F.B., Ross C.W. 1994. Fisiología vegetal. Editorial Grupo Editorial Iberoamericana 759p.

Searle KR, Shipley LA. 2008. The comparative feeding behaviour of large browsing and 382 grazing herbivores. In: Gordon IJ, Prins HHT (eds), The ecology of browsing and 383 grazing. Berlin: Springer Verlag. pp 117-148.

Setter, T.L. 1993. Assimilate allocation in response to water deficit stress. In D.R. Buxton, R.M. Shibles, R.A. Forsberg, B.L. Blad, K.H. Asay, G.M. Paulsen, and R.F. Wilson (ed.) International crop science I. CSSA, Madison, WI. In press.

Silva, L.P., Silva, L.S.W. and Bohnen, H. 2005. Cell wall components and in vitro digestibility of rice (Oryza sativa) straw with different silicon concentration. Ciência Rural 35: 1205-1210.

Skerman, P.J., Riveros, F. 1992. Gramíneas Tropicales. Colección FAO: Producción y Protección Vegetal. No. 23. pp. 257-261

Soderstrom, T.R., Hilu, K.W., Campbell, C.S., Barkworth, M.E. 1987. Grass Systematics and Evolution. Smithsonian Institution Press, Washington, D.C.

Spears, W.J. 1994. Minerals in forage. Fahey, C.G. jr. (ed.). Forage Quality, Evaluation and Utilization. pp: 281-3117. National Conference on Forage Quality. Univ. of Nebraska. Lincoln, Nebraska, E.U. pp.

Stakelumn, G. y Dillon, P. 1991. Influence os sward structure and digestibility on the intake and performance of lactating and growing cattle. En: Mayne, C.S. (editor), Management Issues for the British Grassland

Farmer in the 1990s. Occasional Symposium of the British Grassland Society No. 25. Hurley, pp. 30-42.

Staples, C.R., Fernando, R.L., Fahey, G.C., Jr., Berger, L.L. y Jaster, E.H. 1984. Effects of imtake of a mixed diet by dairy steers on digestion events. J. Dairy Sci. 67:905-1006.

Stern, M. D., and L.D. Satter. 1982. In vivo estimation of protein degradability in the rumen. Protein requirements for cattle: symposium. F. N. Owens, de. Oklahoma State Univ., Stillwater.

Stewart, C.R. y Hanson, A.D. 1980. Proline accumulation as a metabolic response to water stress. p. 173-189. In N.C. Turner and P.J. Kramer (ed.) adaptation of plants to water and high temperature stress. John Wiley and Sons, New York, NY.

Storm, E. y Ørskov, E.R. 1984. The nutritive value of rumen micro-organisms in ruminants. 4. The limiting amino acids of microbial protein in growing sheep determined by a new approach. Br. J. Nutr. 52:613-620.

Struik, P.C., Deinum, B.D., Hoefsloot, J.M.P. 1985. Effects of temperature durig diffferent stages of development on growth and digestibility of forage maize (Zea mays L) Netherlands Journal of Agricultural Science, 33:405-420.

Sutton, J.D., Aston, K., Beever, D.E. y Fisher, W.J. 1992. Body composition and performance or autumn-calving Holsterin-Fresian dairy cows during lactation: food intake, milk constituents output and liveweight. Animal Production, 54: 473-480.

Tamminga, S. 1979. Protein degradation in the forestomachs of ruminants. J. Anim. Sci. 49:1615-1622.

Teller, E., Vanbelle, M. y Kamatali, P. 1993. Chewing behavior and voluntary grass silage intake by cattle. Livestock Production Science, 33: 215-227.

Terashima, N., Fukushima, K.L. He, F. and Takabe, K. 1993. Comprehensive model of the lignified plant cell wall. In: H.G. Jung, D.R. Buxton, R.D.

Hatfield, and J. Ralph (editores). Forage Cell Wall Structure and Digestibility. ASA-CSSA-SSSA, Madison, WI. EUA. p. 247.

Thomas, C. 1987. Factors affecting substitution rates in dairy cows on silage based rations. En: Haresign, L.W. y Cole, D.J.A. (editores) Recent Advances in Animal Nutrition. Butherworth-Heinemann, Oxford, pp. 205-218.

Thompson, F.N. y Studermann, J.A. 1993. Pathophysiology of fescue toxicosis. Agricultural Ecosystems and the Environment, 44: 263-281.

Tisdale, S.L. y W.L; Nelson. 1982. Fertilidad de los suelos y fertilizantes. 1ª edición. Unión Tipografía Hispanoamericana. México. pp. 101-107.

Turner, N.C. y Jones, M.M. 1980. Turgor maintenance by osmotic adjustment: A review and evaluation. p. 87-103. In N.C. Turner and P.J. Kramer (ed.) Adaptation of plants to water and high temperature stress. John Wiley and Sons, New York, NY.

Underwood, E.J. y Suttle, N.F., 1999. The Mineral Nutrition of Livestock, 3a edición. CAB International, Wallingford, p. 600.

Valdez, J.L., L.R. McDowell and M. Koger. 1988. Mineral status and supplementation of grazing beef cattle under tropical conditions in Guatemala. II. Macroelements and animal performance. J. Prod. Agric. 1: 356-361.

Van Soest, P.J. 1993. Cell wall matrix interactions and degradation session synopsis. In: H.G. Jung, D.R. Buxton, R.D. Hatfield, and J. Ralph (Ed.) Forage Cell Wall Structure and Digestibility. p 377. ASA-CSSA-SSSA, Madison, WI.

Van Soest, P.J. 1994. Nutricional Ecology of the Ruminant, 2ª edición, Cornell University Press, Ithaca, NY, Estados Unidos.

Varga, G.A., y Prigge, E.C. 1982. Influence of forage species and level of intake on ruminal turnover rates. J. Anim. Sci. 55:1498-1506.

Varner, J.E., y Lin, L.S. 1989. Plant cell wall architecture. Cell. 56:233-239.

Waggoner, P.E. 1993. Preparing for climate change. In D.R. Buxton, R.M. Shibles, R.A. Forsberg, B.L. Blad, K.H. Asay, G.M. Paulsen, and R.F. Wilson (ed.) International crop science I. CSSA, Madison, WI.

Waldo, D. R., y Glenn, B. P. 1984. Comparison of new protein systems for lactating dairy cows. J. Dairy Sci. 67:1115.

Wedig, C.L., Jaster, E.H. y Moore, K.J. 1989. Disappearance of hemicellulosic monosaccharides and alkali-solube phenolic compounds of normal and brown midrib sorghum x sudangrass silages fed to Holstein steers. J. Dairy Sci. 72:112-122.

Welch, J.G. 1982. Ruminantion, particle size and pasaje from the rumen. J. of Anim. Sci. 54: 885-894.

Weston, R.H. 1996. Some aspects of constraintss to forage consumption by ruminants. Australian J. of Agricultural Research, 47: 175-197.

White, B.A., Mackie, R.I. y Doerner, K. C. 1993. Enzymatic hydrolysis of forage cell walls. p. 455-484. In H.G. Jung, D.R. Buxton, R.D. Hatfield, and J. Ralph (ed.) Forage Cell Wall Structure and Digestibility. ASA-CSSA-SSSA, modison, WI.

Whitehead, D.C. 2000. Nutrient Elements in Grasslands. Soil-Plant-Animal relationships. CABI Publishing, Wallingford, Reino Unido. p. 135.

Wilkie. K.C.B. 1985. New perspectives on non-cellulosic cell-wall polysaccharides (chemial celluloses and pectic substances) of land plants. p. 1-38. En C.T. Brett and J.R. Hillman (ed.) Biochemistry of Plant Cell Walls. Cambridge University Press, Cambridge.

Wilson, J.R. 1982. Environmental and nutritional factors affecting herbage quality. p. 111-131. In J.B. Hacker (ed.) Nutritional limits to animal production from pastures. CAB, Farnham, U.K.

Wilson, J.R. 1990. Influence of plant anatomy on digestion of fiber breakdown. En D.E. Akin, L.G. Ljungdhal, J.R. Wilson y P.J. Harris (editores). Microbial and Plant Opportunities to Improve Hemicellulose

Utilization by Ruminants. Elsevier Sci. Publishing Co., Inc., NY. Estados Unidos. pp. 99-117.

Wilson, J.R. y Kennedy, P.M. 1996. Plant and animal constraints to voluntary feed intake associated with fiber characteristics and particle breakdown and passage in ruminants. Australian J. of Agricultural Research, 47: 199-225.

Wilson, J.R. y Wong, C.C. 1982. Effects. of shade on some factors influencing nutritive quality of green panic and siratro pastures. Aust. J. Agric. Res. 33:937-950.

Wilson, J.R., Deinum, B. y Engels, F.M. 1991. Temperature effects on anatomy and digestibility of leaf and stem of tropical and temperate forage species. Neth. J. Agric. Sci. 39:31-48

Wilson, J.R., y Minson, D.J. 1980 Prospects for improving the digestibility and intake of tropical grasses. Trop. Grassl. 14;2253-257.

Windham, W.R. Barton, F.E., y Himmelsbach, D.S. 1983. High-pressure liquid chromatographic analysis of component sugars in neutral detergent fiber for representative warm-and cool-season grasses. J. Agric.Food Chem. 31: 471-475.

Yokohama, M.T. 1988. Microbiology of the Rumen and Intestine. En: The ruminant animal, Digestive Physiology and Nutrition. D.C. Church (editor). Waveland Press, Inc. Illinois, Estados Unidos. pp. 125-144.

Xing XR, Zhang L. 1998. Review of the studies on silicon nutrition of plants. Chinese Bulletin of Botany, 15(2): 33-40.

VITA – ROQUE G. RAMÍREZ-LOZANO

Address:
Universidad Autónoma de Nuevo León, Facultad de Ciencias Biológicas, Alimentos. Ave. Pedro de Alba y Manuel Barragán S/N, Ciudad Universitaria, San Nicolás de los Garza, Nuevo León, 66455, México.

Education:
B.Sc. in Agriculture, Universidad Autonoma de Nuevo Leon in 1972; M.Sc. in Animal Science, New Mexico State University in 1983; Ph.D. in Animal Nutrition, New Mexico State University in 1985.

Present Position:
Professor, Department of Food Sciences.

Job Duties:
Professor-researcher. Undergraduate and graduate teaching in animal nutrition, statistics and research techniques.

Research Awards:
Sistema Nacional de Investigadores, Nivel Tres

Graduate Student Supervision:
101 undergraduate students, 14 M.Sc. students and 24 Ph.D. students.

Publications:
147 peer-reviewed journal articles, 35 invited papers, 11 books, 7 book chapters, 62 published proceedings articles and 60 abstracts.

Published works have been cited 1483 times as of August 2015, Index h 21 and index i10 55.

Editorial and Professional Service:
Journal of Animal Science, International Goat Association, several Mexican Animal Science Associations

Printed in the United States
By Bookmasters